CONSIDER THE PLATYPUS

CONSIDER
THE PLATYPUS

EVOLUTION THROUGH
BIOLOGY'S *most baffling* BEASTS

MAGGIE RYAN SANDFORD

illustrations by Rodica Prato

BLACK DOG
& LEVENTHAL
PUBLISHERS
NEW YORK

To all who persist.

Text copyright © 2019 by Maggie Ryan Sandford
Illustrations copyright © 2019 by Rodica Prato
Cover design by Katie Benezra
Cover art by Rodica Prato
Cover copyright © 2019 by Hachette Book Group, Inc.

Black Dog & Leventhal Publishers
Hachette Book Group
1290 Avenue of the Americas
New York, NY 10104
www.hachettebookgroup.com
www.blackdogandleventhal.com

First Edition: August 2019

Black Dog & Leventhal Publishers is an imprint of Perseus Books, LLC, a subsidiary of Hachette Book Group, Inc.
The Black Dog & Leventhal Publishers name and logo are trademarks of Hachette Book Group, Inc.

The publisher is not responsible for websites (or their content) that are not owned by the publisher. The Hachette Speakers Bureau provides a wide range of authors for speaking events. To find out more, go to *www.HachetteSpeakersBureau.com* or call (866) 376-6591.

Print book interior design by Red Herring Design
Library of Congress Control Number: 2018962443
ISBNs: 978-0-316-41839-3 (paper over board); 978-0-7624-9363-0 (ebook)
Printed in China

1010

10 9 8 7 6 5 4 3 2 1

CONTENTS

PREFACE: THE LOADED WORD **10**

WHO YOU CALLIN' AN ANIMAL? **12**

DARWIN, DARWIN, *DARWIN...*! **13**

THE ~~TREE~~ RIVER OF LIFE **16**

PART 1: OUTSIDE IN

20
DUCK-BILLED PLATYPUS *(Ornithorhynchus anatinus):*
Something for Everybody

28
POLAR BEAR *(Ursus maritimus):*
Black and White and Read All Over

34
BLUE WHALE *(Balaenoptera musculus):*
A Huge Problem for Darwin?

42
TRINIDADIAN GUPPY *(Poecilia reticulata):*
Teacher's Pet

46
MASAI GIRAFFE *(Giraffa camelopardalis tippelskirchi):*
It's Not All in the Neck

52
HORSE *(Equus caballus):*
Born to Run

56
PEPPERED MOTH *(Biston betularia):*
Want to Paint It Black

62
DAPHNIA *(Daphnia pulex)* AND NEMATODE *(Caenorhabditis elegans)*

64
RED FLOUR BEETLE *(Tribolium castaneum):*
Cutting Out the Middleman

68
GIANT MARINE ISOPOD *(Bathynomus giganteus):*
Easy There, Big Guy

70
HUMAN *(Homo sapiens):*
Definitely Not Descended from Monkeys

76
NEANDERTHAL *(Homo neanderthalensis):*
Actually People

84
BOTTLENOSE DOLPHIN *(Tursiops truncatus):*
All in Their Heads

88
MANATEE *(Trichechus manatus):*
Curves in All the Right Places

90
AMERICAN CROW *(Corvus brachyrhynchos):*
"Flying Monkeys"

94
AMERICAN PLAINS BISON *(Bison bison)* versus COW *(Bos taurus):*
Son of a Beefalo

100
NEAVES WHIPTAIL LIZARD *(Aspidoscelis neavesi):*
Virgin Births

106
HOATZIN *(Opisthocomus hoazin):*
Stuck in That Awkward Phase

110
AFRICAN MALARIA MOSQUITO *(Anopheles gambiae):*
Why, Though?

114
DOMESTIC DOG *(Canis lupus familiaris):*
Weirdo Wolves with Humans for a Pack

124
CAT *(Felis silvestris catus):*
Living Its Best (Double) Life

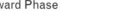

PART 2: SONG OF MYSELF

136
FRUIT FLY *(Drosophila melanogaster):*
The Original Supermodel of the World

144
VENT SHRIMP *(Rimicaris exoculata):*
Eyeless Wonder

146
GREAT BARRIER REEF DEMOSPONGE *(Amphimedon queenslandica):*
Your Simple Cousin

152
COMB JELLY *(Mnemiopsis leidyi):*
Yin and Yang—Some of This, None of That

156
SOUTH AMERICAN LUNGFISH *(Lepidosiren paradoxa)* versus
AFRICAN COELACANTH *(Latimeria chalumnae)*
Fossil or Fam?

162
ZEBRAFISH *(Danio rerio):*
Full Transparency

164
AFRICAN CLAWED FROG *(Xenopus tropicalis, Xenopus laevis):*
Sleeper Cells

168
ZEBRA FINCH *(Taeniopygia guttata):*
Talking the Walk

172
GRAY SHORT-TAILED OPOSSUM *(Monodelphis domestica):*
Great Grandmammal

176
NINE-BANDED ARMADILLO *(Dasypus novemcinctus)* and
HOFFMANN'S TWO-TOED SLOTH *(Choloepus hoffmanni):*
Cousins from Way Back

182
HOUSE MOUSE *(Mus musculus):*
Of Mice and Men

186
BONOBO *(Pan paniscus):*
Other Brother from Another Mother

PART 3: HOPELESS MONSTERS

192
JUNGLE FOWL, AKA DOMESTIC CHICKEN *(Gallus gallus):*
Which Came First? The Chicken

196
GHARIAL *(Gavialis gangeticus):*
Bird Brain

200
BARN OWL *(Tyto alba):*
Surround Sound without a Sound

202
VELVET SPIDER *(Eresus kollari):*
The Elements of Style

208
AYE-AYE *(Daubentonia madagascariensis):*
Perhaps Our Least Cute Nearish Relative

212
NORTHERN BROWN KIWI *(Apteryx mantelli):*
The Mammal of Birds

216
GALAPAGOS TORTOISE *(Chelonoidis nigra):*
Enter the Hopeful Monster

222
LITTLE BROWN BAT *(Myotis lucifugus):*
Flying under the Radar

PART 4: THE SECRET TO ETERNAL LIFE

230

AXOLOTL *(Ambystoma mexicanum):*
Forever Young

234

NAKED MOLE RAT *(Heterocephalus glaber):*
Only as Old as You Feel

238

HONEYBEE *(Apis mellifera):*
Hive Mind

242

AFRICAN ELEPHANT *(Loxodonta africana)* versus
WOOLLY MAMMOTH *(Mammuthus primigenius):*
Death Match

246

TARDIGRADE *(Hypsibius dujardini):*
Persistent Little Buggers

254

CALIFORNIA TWO-SPOT OCTOPUS *(Octopus bimaculoides):*
Brother from Another Planet

AFTERWORD **264**

SPECIAL THANKS **265**

INDEX **266**

ABOUT THE AUTHOR **272**

PREFACE: THE LOADED WORD

When your passion is communicating with the public about science, you'll talk about your work to anyone who will listen, and listen to anyone who will talk. You start to notice patterns in people's reactions to certain topics. *Evolution*, it turns out, is a word that gets a reaction, which can range from name-checking terms like "survival of the fittest" to jokes about being a monkey's uncle, to starting debate about theology.

Evolution is a weighty topic, with as many data points as the number of cells that are alive on Earth. Only biologists who really specialize in the subject are willing to lean in to conversation about it, as it is so fraught with core beliefs.

Take, for instance, this anecdote by the evolutionary biologist Jonathan B. Losos, from his book *Improbable Destinies.* He describes a conversation he had on an airplane, en route to conduct a field study on the evolution of color in desert mice, for which he invented a special fencing technique. When the gentleman in the seat next to him asked about his work, he happily described the experiment. His fellow traveler had grown up on a farm, so he was familiar with animal proliferation—that's what breeding livestock is all about, after all. But as soon as Losos let the name Darwin fall from his lips, the mood of the conversation nosedived. When they were talking mice and mating and coat color and fencing, the two men were speaking the same language. But when animal husbandry became *evolution*, it became a loaded word.

Sometimes the best way to make a concept less weighty is through a story, such as Red Riding Hood's cute cautionary tale as a stand-in

for the harsh risks of talking to strangers. The stories here belong to the animals. Or rather, they belong to entire families of animals, lineages, their arcs told in geological time.

In selecting stories for this book, I attempted to be democratic in my sampling, to include animals from far-flung corners of the animal kingdom, animals beloved and reviled and rarely heard of. Animals that have been scantly researched appear alongside heavily researched animals—those "greatest hits" animals that show up in every evolution textbook.

I grilled evolutionary biologists for their favorite critters and plowed through texts for standouts. I stood on the shoulders of those who went before to see what they'd seen and hadn't seen, then went to Google Scholar to see if anyone had seen anything lately and how many people had cited them for it. I thought I'd settled on about 140 animals at one point, then painstakingly whittled the list down to the menagerie you see here.

Together, I hope they help illuminate some of the vast, deep, weighty, loaded story of evolution. If nothing else, the experts I talked to seemed to be glad I was taking on this effort

Beagle Laid Ashore, River Santa Cruz, 1839. By Conrad Martens, a landscape artist who traveled alongside Charles Darwin on the **HMS** *Beagle*'s fateful second worldwide voyage.

instead of them, so heyyyy! Happy to help! Even though this project sometimes made me feel like Red Riding Hood herself, venturing one step at a time into a big dark woods with way too many paths and a basket not nearly big enough for everything I needed to carry. It was worth it though to meet the creatures that teemed in the branches and streams and under the dirt.

WHO YOU CALLIN' AN ANIMAL?

This is a book about animals, by animals, *for* animals. I'm an animal. And so are you.

If you have an adverse reaction to being called an animal, remember this: it's just a word, a word made up by animals to describe other animals, only to later realize that we're among them.

For the purpose of this book, though, I use the definition of *animal* that is currently universally accepted by the scientific community: an organism (living thing) that:

- is made up of more than one cell (multicellular)

- feeds on organic matter

- rapidly responds to stimuli

- reproduces

In short: something that is alive but is not a plant, fungus, virus, bacteria, or other single-celled thing. If you're disappointed to find your favorite animal missing from this book, let me tell you: me too. (The scarcity of birds profiled is downright criminal.)

Overall, though, the choice of featured animals herein represents a microcosm of the study of animal evolution.

WHAT IS THE STUDY OF EVOLUTION CALLED?

Not even that has a single name. Evolutionary biology is the most common, but the researchers can be paleontologists, ecologists, zoologists, taxonomists, physiologists, behavioral neuroscientists, embryologists, oncologists, and now geneticists. Researchers can even be laypeople committed to counting the ticks on their dogs or crows in their yards each season.

Like the process of evolution itself, the process of understanding it is messy. It started blind, crawled along from one fossilized tooth to another, one dissection, one fuzzy DNA gel, one scattershot genome to the next, until slowly those patterns started to emerge.

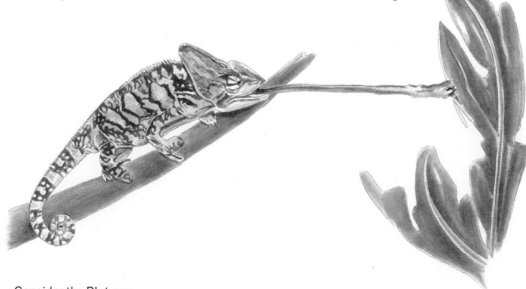

DARWIN, DARWIN, *DARWIN*...!

You'll be seeing a lot of Charles Darwin in this book. He's credited with the theory of evolution, and although he wasn't the only one onto it, he really kicked things off. His isn't the only voice that has contributed to our understanding of evolution, but his was the primary, foundational voice. Think of his presence here as a case study in evolutionary biology, much in the same way each animal here is a case study.

By the end of his life and long after, Darwin is a key figure in the study of evolution, but he wasn't always a formidable genius.

His evolutionary story, if you will, began at age 16, when young Charlie escaped the relative drudgery of his medical apprenticeship to learn from one John Edmonstone, a freed slave who taught taxidermy at the University of Edinburgh.

At age 22, much to the dismay of his wealthy doctor father, Charles snuck off to board the HMS *Beagle*, having accepted relatively meager wages to be resident naturalist on what was to be a two-year voyage charting the coastline of South America. By the time he

Here he is as a child with his sister, holding a potted plant like a dork.

Here he is as a young man with an awkward hairdo.

returned to London, Darwin had been away four years, circumnavigated the globe, and observed a lifetime's worth of biodiversity. This journey sowed the mental seeds that would become his theory of evolution.

It took Darwin 20 years after his return on the *Beagle* to take his ideas public. He might never have gotten up the gumption had it not been for Alfred Russel Wallace, a young naturalist and admirer of Darwin's who wrote him a series of letters about a nascent theory of his own. Fueled by passion and friendly competition, the two joined forces and published as soon as they could. Though Darwin claimed the legacy, Wallace remained a key collaborator and eventually became his friend.

Darwin's theory laid the foundation upon which all modern evolutionary scientists have

THEORIES ON DARWIN'S THEORY

One piece of jargon important to call out is the word *theory*. In casual conversation, a theory is just a theory—something you're mulling over and don't yet have proof to substantiate. But when academics use the word *theory*, they mean something completely different.

Theory in this case is a school of thought, an idea framework, a way of organizing ideas that apply to your discipline and that other people in your discipline might find useful. Gravity, technically, is a theory. And yet here we are not flying off into outer space

because this theory is the most logical explanation about why things happen as they do.

This isn't to say that theories can't be built upon. That is, in fact, what they're for. Albert Einstein couldn't have "corrected" Isaac Newton's various theories of physics if Newton hadn't laid them out—Einstein probably never would have been a physicist in the first place. But theories don't get any traction unless they've earned it by holding true, for a lot of people, for a long time. So ultimately, this is just a problem of semantics.

continued to build. It holds up even today under the ever increasing informational mass of fields like genetics and genomics, despite the fact that the gene itself—the key mechanism for heritability—was so elusive to Darwin that he nearly suffered a mental breakdown in its pursuit.

Beyond this, you won't see too many names of other scientists in this book, as I don't want you to get sidetracked by them. Or by dates. Or by a lot of jargon, which is necessary for science, but it's a double-edged sword. If you understand the jargon, you've got an automatic "in"; if you don't, it might drive you away.

A Venerable Orang-outang, 1871. One of many caricatures of Charles Darwin, this from a satirical magazine printed two whole years after his treatise, *On the Origin of Species,* first dropped.

THE ~~TREE~~ RIVER OF LIFE

The notion of a family tree is a comfortable one, a useful tool for one to trace one's proud parentage back through the ages. It's a logical leap to extend the tool to connect all life on Earth, an image that has emerged repeatedly in the minds of naturalists even pre-dating Darwin. But Darwin made it famous. For 32 years, as he worked through the ideas that would become his theory, he sketched "trees of life" over and over, most especially in a private, back-burner notebook he'd simply labeled "B."

One particular tree from this notebook has become iconic and is sometimes hailed as his eureka moment. It depicts a single ancestor (1) begetting many others who beget others and eventually we end up with living species (A on one branch; B, C, and D on another). You'll see it posted on laboratory walls and classrooms—even tattooed on the bodies of true evolution nerds.

But while this is one of the cleaner and more complete versions of Darwin's tree, it was one of his first. It dissatisfied him. (Note the "I think" scrawled above it, as if to ward off the ill luck of hubris, even in his own private journals.) It was incomplete. For starters, the tree itself is alive, which is sort of like using a word to define itself. On another page he considers instead another option, ruminating:

"The tree of life should perhaps be called the coral of life, base of branches dead; so that passages cannot be seen."

But later he decides:

"No only makes it excessively complicated… contradiction to constant succession of germs in progress." (By "germs" he meant genes, or what would eventually come to be called genes. Another term he hunted for all his life.)

Naturalists before Darwin and many more after him have all tried to reimagine the tree of life, up to and including a massive cross-institutional effort published in a 2016 article in the scientific journal *Nature*, that allowed the tree to spiral in on itself so as to accommodate all the information springing from it. In the two years since, the explosion of genomic science has begun to work with the image of a web of life, acknowledging connections shared by species that haven't shared an ancestor for millennia or more.

For the purposes of this book, consider a river.

Like a tree, a river branches, but its path is determined only by its environment and the laws of physics (such as gravity, friction, and motion). It does not have a predetermined shape; it doesn't have a driving "life" force. It simply runs toward the ocean, diverted by rocks, hills, valleys, and weather. Over time and space, some rivulets dry into gas and reenter the water cycle (like animals dying off, ending lineages and returning organic molecules to the environment), while other rivulets might split, shrink, or grow.

The river has volume, too, three dimensions rather than single points linked by lines. When we talk about "an animal" here we're not (usually) talking about a single animal from which all others spring. We're not even talking about two animals and their lineage, like the "begats" on a human family tree. We're talking about groups of animals, populations over time, that bred and changed en masse, branched and met back up, grew and shrank, left a few fossils behind and brought traces of all of those changes with them.

It's much easier to picture the complexity and sheer amount of information in the mix this way, a flooded gene pool in motion. The weight of the information we've learned (and have yet to learn) from the study of genomes has become too great for a tree branch to hold. A river can always grow wider and deeper.

In the larger context, the headwaters become the place where life begins. Imagine organic molecules coming together for the first time, like how water exists in the water table and soil and as water vapor in the atmosphere, but it doesn't become a river (life) until it burbles forth or rains down into the riverbed. It only becomes a river (life) when it gains momentum and runs. It is life as long as it is moving forward.

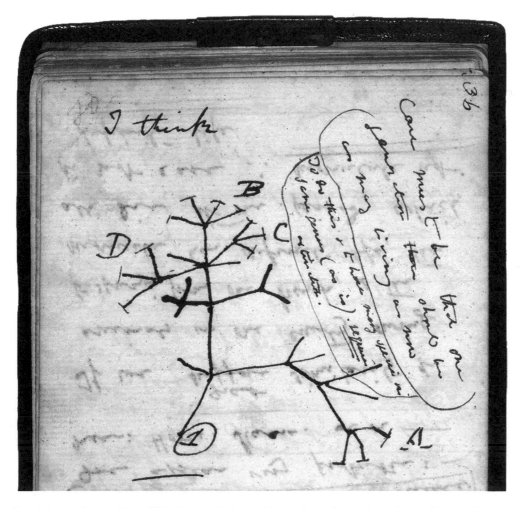

Darwin's most famous Tree of Life diagram, just one of dozens he worked and reworked in his journals.

PART 1:
OUTSIDE IN

Kids today, with their 23andMe and Ancestry.com, might not remember a time when just *observing* was the only way to determine what something was and where it might have come from.

Before evolution was a theory, it was a bunch of wild animals doing what wild animals do and a bunch of scientists observing them and postulating. Observation was the methodology by which Darwin developed his theory, which he then presented in a publication whose full title breaks it down for you: *The Origin of Species by Means of Natural Selection or the Preservation of Favoured Races in the Struggle for Life*. In this section, we'll unpack these phrases in the context of evolutionary biology:

■ Species

■ Natural Selection

■ Preservation of Favoured Race in the Struggle for Life
   (which we'll just call "Success")

In this section, we'll apply them to animal appearances. And we'll hold Darwin's theory up to the scrutiny of new genetic understanding.

DUCK-BILLED PLATYPUS

(Ornithorhynchus anatinus)

SOMETHING FOR EVERYBODY

Platypuses have long been considered oddballs in the animal kingdom, which makes them the perfect first stop on our journey into the weird world of evolutionary science.

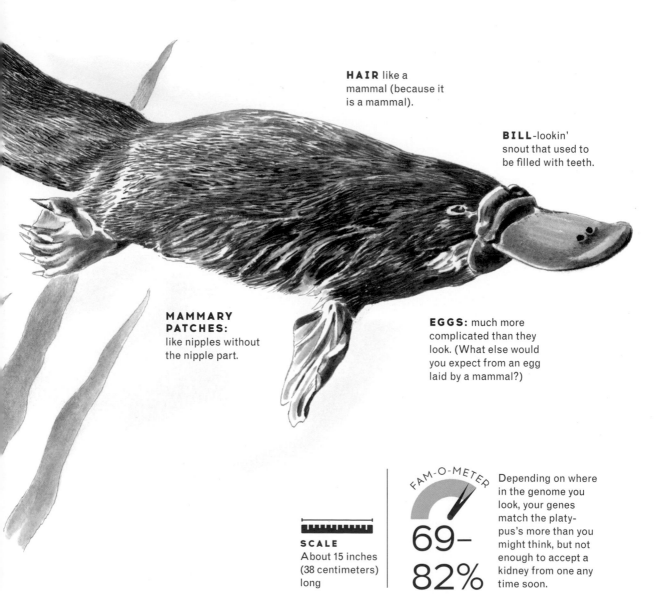

HAIR like a mammal (because it is a mammal).

BILL-lookin' snout that used to be filled with teeth.

MAMMARY PATCHES: like nipples without the nipple part.

EGGS: much more complicated than they look. (What else would you expect from an egg laid by a mammal?)

SCALE
About 15 inches (38 centimeters) long

FAM-O-METER
69– 82%
Depending on where in the genome you look, your genes match the platypus's more than you might think, but not enough to accept a kidney from one any time soon.

Having diverged from the rest of Animalia just after the emergence of modern mammals, the platypus finds itself at the tip of lonely rivulets off the "Monotreme Stream" next to echidna like *Zaglossus attenboroughi* (named for the beloved television naturalist). Funny thing about evolutionary outliers, their ancient relatives look just like them. "Living fossil" is of course a misnomer (for more, see Coelacanth, page 157), but a nicer way to describe animals that time forgot might be "If it ain't broke, don't fix it." Since the platypus's isolated Australian habitat has changed only slightly through the millennia, the only thing that time forgot was about two feet and its teeth. Prehistoric proto-platypuses included three species of oversized, carnivorous proto-platypuses, which make the whole family line seem less silly and more kill-y. But only a little.

PLURALITY

Before you drop this book to start writing the publisher about an error, the official plural for "platypus" is in fact "platypuses," not "platypi"—sad news too for fans of octopuses and cactuses. (If you want to be real snotty, you can go with the OG Greek *platypodes*.) What's a group of platypuses called? Some say string (boring), some say puddle, but the best/most official is a paradox of platypuses. You'll soon see why. Hint: It has to do with plurality too. The newest way to think about evolution is in plurals: there's more than one story in every animal, and more than one animal in every story.

WHEN IN (PRE)HISTORY/ WHERE ON THE RIVER

360	250	MILLION YEARS AGO	65	0
PALAEOZOIC		MESOZOIC		CENOZOIC

...Split from the branch that became the rest of the mammals, including humans, about 166 million years ago

THE PLATYPUS LINEAGE

...Split from the branch that would eventually become modern reptiles and birds 315 million years ago

...Split from its closest extant relative, the echidna (aka spiny anteater), 17–80 million years ago

Charles Darwin caught a fleeting glimpse of a platypus when the *Beagle* moored in Australia for two months at the very end of its four-year voyage around the world.

He journaled:

In the dusk of the evening I took a stroll along a chain of ponds, which in this dry country represented the course of a river, and had the good fortune to see several of the famous Ornithorhynchus paradoxus. They were diving and playing about the surface of the water, but showed so little of their bodies, that they might easily have been mistaken for water-rats. Mr. Browne shot one.

Via notebooks and correspondence in the years to follow, Darwin often expressed concerns that "the strange and inexplicable fact of *Ornithorhynchus*" was so perplexing in its taxonomy that it alone might prove to be the undoing of his nascent Theory.

He also wondered: "When will *Ornithorhynchus* come in circle?...Such difficulties will always occur if animals are thought to have been created out of nothing, by God."

BEASTLY BREAKDOWN

HAIR

It's Europe in the late 1790s. The term *scientist* won't be a thing for 35 years. But the folks who study the natural world know so little about the platypus that when they get hold of their first study skin, they think it's a fake. (Faux exotic animals are in vogue at this time, like the infamous Feejee Mermaid, aka half a dead monkey with a dried fish tail sewn on.) Science in the 1790s says that fur = mammal, but this one is like no other mammal they've seen. They spend the next 200 years trying to decide how to classify platypuses, poring over scant specimens, arguing among themselves, and often putting more stock in their own made-up system of classification than in the evidence before their very eyes.

Different groups of researchers arrived at different understandings of the platypus's situation as each group found more information. Others ignored new information and continued believing what they wished to. (Yes, scientists do this, too. But not the good ones.) It wasn't until 2008 that a team of geneticists from Asia, Europe, Australia, New Zealand, and the US all came together to sequence the platypus genome. The genome belonged to Glennie, a female platypus named after the Glenrock area of New South Wales where she was found. But you can think of her as the John

Glenn of platypodes, rocketing monotreme science into the future—and nearly into the realm of science fiction. Patterns in her DNA matched patterns in a baffling array of different genera. Patterns associated with her dense, oily fur, though, were indeed mammalian, specifically reminiscent of genetic fur-making patterns in otters and beavers.

BILL

In 1799 a naturalist named George Shaw documented the freaky beast and its parts, giving it a name that eventually partially sticks: *Platypus anatinus* (from the Greek for "flat-footed" and the Latin for "duck-like").

The bill's lack of teeth misled early naturalists to assume the bill was bird-ish. But in the 20th and 21st centuries, paleontologists studying the platypus's fossil record found teeth from early relatives of the platypus. In 2013, an American paleontologist unearthed a molar from an ancient duck-billed monotreme that was about 3.5 feet (more than 1 meter) long and probably ate large prey like frogs, birds, and entire turtles. She was the first to assert that as these "proto-puses" evolved and shrank, a ridged bill and rough tongue became sufficient to eat smaller food, and it lost its teeth to the ages.

But that's not all: a team of geneticists in 2008 found that Glennie's bill featured an elaborate radar system (sort of): a combination of touch receptors and electroreceptors that allowed her to pick up movements and low-frequency electrical signals in her prey like some kind of Dadaist shark. Indeed, most of the other animals that use electricity as a sixth sense are fish. But the genetic patterns that bring about the trait in platypuses read the same.

EYES

Back in 1779, Shaw agreed with Australian colonists' nickname for the platypus, "water

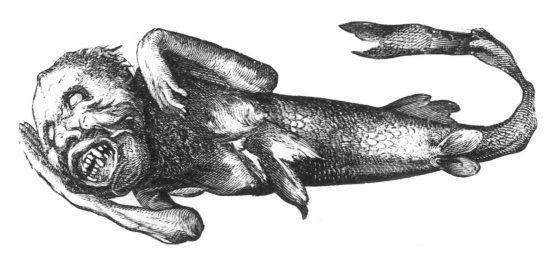

For those of you just looking at the pictures: *This is not a real thing.* The "Feejee Mermaid" was faked by a taxidermist, but is a useful lesson: (1) ABS: Always Be Skeptical.

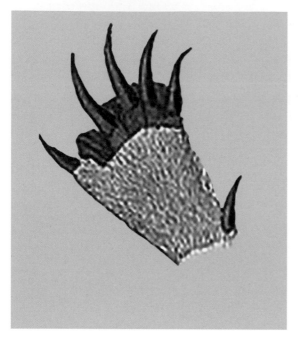

George Shaw himself produced these rather questionable drawings of the platypus's anatomy and in 1799 published them in *The Naturalist's Miscellany, or Coloured Figures of Natural Objects*. Included is the male platypus's venom spur.

mole," suggesting that its beady little eyes seemed as useless as those of the subterranean garden ruiners.

But in 2008, the genomic team findings suggested that the platypus's eyes have a rod/cone balance that most closely resembled that of other mammals. They also have a "double cone" thing going on, a feature found in neither eutherian mammals (mammals that give birth to fully developed young) nor marsupials (who keep their young in pouches for some time after birth —e.g., kangaroos and opossums). Meanwhile, their eyeballs are enclosed by a type of cartilage more like that seen in birds, reptiles, amphibians, sharks, rays, and creepy movie monsters.

FEET

The mole comparison extends beyond platypuses' eyes to their massive clawed feet, which—like moles—they use to dig complex burrows.

And while the 2008 genomic team couldn't say exactly which part of the platypus's DNA codes for "feet," they could point out the genetic markers for their venom. Delivered by a sting from spurs in the male platypus's hind legs, this venom contains genetic codes similar to the venom in reptiles. But it also contains strings of code found in mammals, most of which aren't venomous. A likely explanation is that venom was a trait found in a prehistoric ancestor shared by most extant animal families (birds, reptiles, mammals, fish), but most mammals and birds lost it as they evolved.

There are a few other mammals that are venomous, however, and one of them is a mole-like shrew called the solenodon. Coincidence? Maybe. Co-incidence, yes. Coincidence like "freak accident," no.

EGGS

Eggs are small, humble things. But if you've ever tried to answer the chicken or egg question, you can see how confounding eggs can be for a person just trying to get some answers.

In 18th-century Australia, European colonists started showing up and shooting animals and arguing about how to classify them. The chief of a local indigenous community tried to tell the Europeans that it was common knowledge among his people that female platypuses lay eggs. The eggs are approximately the same size and color of small chicken eggs, laid two at a time, always in a nest among the reeds just atop the surface of the water. He said the motherpus spends a lot of time sitting on the eggs. And by the way, this animal already has a name, mallangong. The Europeans made note of his quaint anecdote but decided they required more proof.

Meanwhile, back in Europe, the young French naturalist Étienne Geoffroy Saint-Hilaire was steadfast. His peers had mostly decided the platypus is a mammal, but he was dead-set on proving them wrong. He had finally gotten his hands on a platypus specimen preserved in alcohol with its insides intact and observed the presence of a cloaca: a posterior orifice found in birds and reptiles that serves as an exit for both

excrement and eggs. That's compared to three separate holes for vagina, urethra, and anus in mammals. His colleagues would find platypus eggs eventually, he insisted in an article, dubbing the animal's theoretical taxonomic order "Monotreme": mono = one, treme = hole. (Saint-Hilaire also quotes the indigenous chief in his article while reassuring the reader that, even though his only ally on the egg issue lacks civilized education, the chief is also bright, insightful, and "lacks neither light nor morality.") In 1844, Saint-Hilaire died still believing that the platypus would one day prove to lay eggs.

It would be 40 more years before Western scientists would settle the matter for themselves, with a now famous telegram from a field zoologist in Australia to Cambridge University: "Monotremes oviparous, ovum meroblastic." (Old-school biologists often used Latin to communicate across language barriers, and also to sound fancy. It's unclear which was the case here.) It meant: Monotremes lay eggs, and the eggs are most similar to those of reptiles.

IN 2008, THE PLATYPUS'S GENOME FURTHER SHOWED THAT ITS EGGS SHARE TRAITS WITH A VARIETY OF ORDERS THROUGHOUT ANIMALIA:

- Like reptilian eggs, platypus eggs are leathery.
- Like in marsupial mammals such as kangaroos, platypus young are born (i.e., hatch from their leathery eggs) in a semi-fetal state, and continue developing throughout the nursing process.
- Platypuses exhibit several genetic markers that are related to egg-making birds, amphibians, and fish.
- Platypuses exhibit several codes found only in egg-making birds and fish.

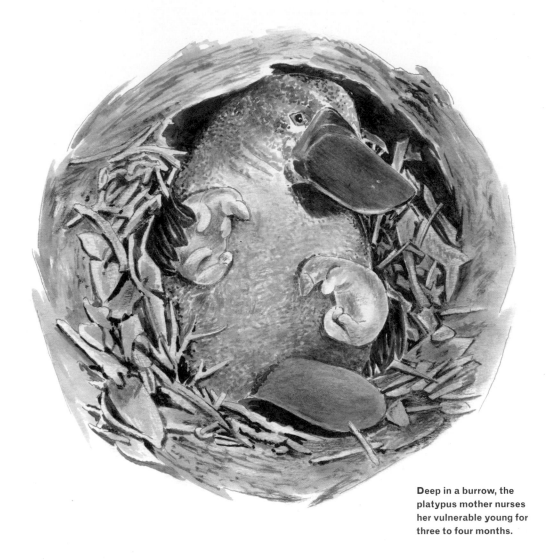

Deep in a burrow, the platypus mother nurses her vulnerable young for three to four months.

MAMMARIES (NIPPLE-LESS)

Early European naturalists long ago defined mammals as creatures that nurse their young on milk produced by mammary glands. But when, by the 1820s, they can find neither platypus nipples nor eggs, they worry that their classifications might fall short of nature's reality. (An academic's worst nightmare.) Saint-Hilaire's ideas came perhaps closest to the eventual truth, but while he insisted that platypus eggs would someday appear, he also insisted that platypus nipples *never* would.

Then in 1824, a young German anatomist discovered a platypus that was producing milk.

Saint-Hilaire was right, in a way: the platypus doesn't have teats. Instead, its milk oozes from mammary "patches" that emerge as milk comes into the mammary glands after it gives birth and basically disappear by the time their young are weaned. This is the first time European scientists have seen such a thing; the only other animals with mammary patches are the other Australian members of the monotreme family, four species of echidna (spiny anteaters). But mammary patches are still, well, mammary (adjective meaning "makes milk"). The platypus caused Western science to expand its definition of mammals and their namesake (mammaries) alike.

The 2008 study concluded that the platypus's genome shows markers for "milk proteins despite eggs," genetic evidence that those 18th- and 19th-century naturalists had good reason to argue. Looking back through the evolutionary timeline, this suggests that the platypus and its cousins broke away from the rest of the furry family right around the same time milk production started being a thing. And milk has been the (r)udder guiding mammal evolution ever since.

Still…The more we learn about genomes, the more we realize that the platypus isn't so strange after all. Glennie just wore her weirdness on the outside.

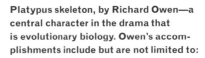

LESSON OF THE PLATYPUS:

Don't believe everything you see. Categories like "mammal," "reptile," and "bird" might not hold up the way they used to…if they ever really did.

TERMS DEFINED: Mammal

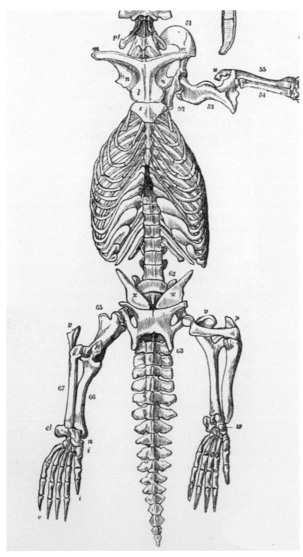

Platypus skeleton, by Richard Owen—a central character in the drama that is evolutionary biology. Owen's accomplishments include but are not limited to:

- coining the terms "dinosaur" and "protozoa"

- pioneering the natural history museum as an institution

- shrilly criticizing his young colleague **Darwin.**

Owen's work made Darwin's better, though, and Owen raised questions that modern genomic science has only just begun to answer.

NECK: long and flexible for hunting blubber balls (aka seals)

HAIR: white and waterproof; each strand has a hollow core

HEAD: that's streamlined for swimming but gets broader the more grizzly genes it has

FEET: covered in fur, even between the toes

SCALE
Female: 6.6 feet (201 centimeters) tall, male: 4.4 feet (134 centimeters) tall

FAM-O-METER

80%?

The exact human/polar bear genetic overlap has yet to be calculated, so the estimate here draws from the human/dog overlap, the bear's close-ish relatives.

POLAR BEAR

(Ursus maritimus)

BLACK AND WHITE AND READ ALL OVER

The polar bear, with its characteristic white coat, stands alone in bear-kind. Or does it? A unique animal uniquely adapted to a disappearing habitat, the polar bear has become a poster child for climate change awareness. But now, thanks to recent genetic findings, it may well become a poster child for skepticism and not taking a single word (like *species* or *polar bear*) for granted.

For many years, the scientific community accepted that polar bears emerged between and 130,000 and 600,000 years ago. The oldest polar bear fossil humans have unearthed dates to about 115,000 years ago, so it had to be before that.

But then in 2010, genetic and dental analysis of a 150,000-year-old fossil pegged its origin as part brown bear and part polar bear. The bear science community fell into sheer chaos.

Based on what they know already, the bear scientists who made the discovery think the ancient hybrid emerged from early brown bears. The fossil looks like a transitional fossil, so they thought it must be a time capsule of the very moment polar bears began to emerge. They put out a press release. "Polar Bears Just 150,000 Years Old," the headline announced.

Yes and no.

In 2012, a different study looked at certain sections of DNA from a different ancient bear tooth, and announced finding crossover in DNA that dates polar bears as being closer to 600,000 years old. "Polar Bears 450,000 Years Older than Previously Thought," another headline announced, pointedly.

Yes and no.

Meanwhile, the first headline is still out on the internet. Currently, there's no correction or update. At the moment the article was published, that was the closest thing to the truth.

There are a lot of questions, gaps, and complications of wording here. For one thing, just because a brown bear is brown doesn't mean that it's a "brown bear" in the genetic sense. Brown bear is another name for modern-day grizzly bears. They were so named by the explorers Lewis and Clark because their coat is sometimes a grayish, blondish brown, like the beard of a grizzled old man. Before they settled on the name, they'd journaled about "white bears," "golden bears," and "brown bears" before finally taking some pelts to a Native hunter who told them they were all the same bear. Grizzlies are different colors at different times of the year, sure. But maybe there was more to the color variation than Lewis and Clark could have known.

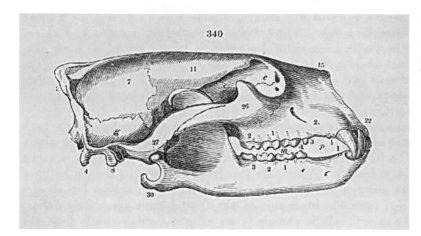

From Richard Owen's *Anatomy of Vertebrates*, **1866. Compare the blunt skull of the grizzly bear to the longer skull of the polar bear on page 32.**

SPECIES: WHAT'S IN A NAME?

Species is a term that has changed definition through the history of biological science. The term originates in the ancient Latin *species*, meaning "a particular sort, kind, or type" or "a sight, look, view, appearance." Carl Linnaeus, an 18th-century naturalist and physician, applied the term to plants and animals in the mid-1700s. That was back when he was shaping early biology and busy naming a lot of things in Latin, including starting the system of universal scientific names, which he thought should also be in Latin, because what could be more universal than that?

The more we learned about the relationships among animal groups, however, the more it seemed that an animal's species could be as difficult to determine as the genus of a platypus. In 1942, evolutionary biology pioneer Ernst Mayr defined *species* as a group of animals that can interbreed and produce healthy offspring. That definition stuck for many decades and is still widely accepted today. But the times they are a-changing. The more we learn about the differences and similarities between and among animals, the more the definition and application of the term *species* is subject to change. It is still useful as an organizational tool, but many researchers find it outdated, a remnant reference that our collective scientific knowledge has since outgrown. The word *species* remains one of the most hotly debated topics in evolutionary biology today.

Hybrid: Species + Species = ?

Intertwined with the term *species*, the term *hybrid* has been around since the 17th century and has traditionally referred to offspring of two separate species. By definition, this type of hybrid is assumed to be infertile, or "inviable." But if the hybrid can itself reproduce, the term "viable hybrid" is called in to bridge the gap. Here again, genetic technology seems to have outpaced the terminology.

WHEN IN (PRE)HISTORY/ WHERE ON THE RIVER

65 MILLION YEARS AGO
...during that major mammalian split, the lineage that would come to include bears emerged along the carnivore path.

30 MILLION YEARS AGO
...it split from cats and hyenas to go the way of the dog and walrus.

25 MILLION YEARS AGO
...there was bear. The polar/grizzly/black bear lineage split from the lineage that would itself split into lineages that separated Asian black bears from spectacled bears and sloth bears (not actually sloths, barely bears).

3.5 MILLION YEARS AGO
...the bear lineage split from the lineage that would eventually lead to panda "bears." This doesn't mean they're not bears; it just means that bears are a lot more varied than we think and (here again) humans just name animals as we go. There is no universal definition of *bear* handed down from on high.

3.3 MILLION YEARS AGO
...the polar bear/grizzly bear lineage split from the lineage that would become American black bears.

1-1.5 MILLION YEARS AGO
...the common ancestor of the brown bear and the polar bear diverged from the rest. But recently, the exact timing of the modern polar bear's divergence has been the source of some contention.

ABS, Always Be Skeptical: Grizzly bears eat a lot of things: roots, grasses, berries, pine nuts, rodents, fish, insects. They may go after larger prey like young elk or even black bears, but not as exclusively as their brothers to the north. This 1895 depiction plays up the drama for effect. Or maybe this grizzly had more than a little polar bear in him.

BEASTLY BREAKDOWN

Scientists have long known that hybrid grizzly polar bears do exist, with the terrific name grolars. But now we can dig into their genomes. There's an interesting opportunity afforded by the chance to examine the exchange of genes between separate "species" of brown bears and polar bears. Most people, by the way, just say "mate," not "exchange genes," but scientists know that mating is just a formality: the real thrill is in finding out where all those exchanged genes end up and how they express themselves. Scientists call this phenomenon **gene flow.**

HEAD

A polar bear's head is more slender than a grizzly's, which presumably helps it stay hydrodynamic in cold waters (not unlike a whale or a dolphin). A polar bear/grizzly hybrid's head shape is somewhere in between. This reminds us that multiple genes contribute to head shape.

METABOLISM

Brown bears can survive in a variety of habitats (generalists, you'd call them). But polar bears evolved to take advantage of a very specific ecological niche: dining on the high-fat meat of a few specific species of arctic seal. It takes a lot of stored and expendable fat to swim around among ice floes. As such, the polar bear has evolved metabolic limitations, much like we'll see in the domestic cat: there's no reason to adapt to eating plants in a world where nothing grows.

But as arctic ice melts, polar bears do more swimming and finding less to eat–nearly half the bears in a 2018 study didn't catch enough food to sustain their daily activity. The unfortunate bears were forced to either scavenge carcasses or go without. These animals lost 10% of their body mass over about 10 days. Grizzly genes could help them adapt to their changing environment: like the ability to eat whatever they want and save up body fat like grizzlies do for their long winter hibernation.

SIZE

Grizzlies are about a foot taller than polar bears on average. Many of a grolar's traits defer to looking like one or the other of its parents, but overall size and shape showed up as a combination of the two.

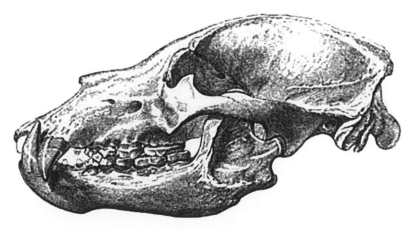

**Polar bear skull.
Richard Owen, 1846.**

TAIL

Polar bears have visible tails. Grizzlies do not. Grolars, so far as we've seen, always have a visible tail. This is really just a function of how the tail lies in the fur, but since a polar bear's fur game is a lot tighter (literally: its fur has to seal up more tightly to keep the bear warm and water-resistant), this suggests that some of the polar bear's useful cold-battling fur traits have remained dominant in the gene pool.

HAIR

If you look at a cross section of the hair of a polar bear, you'll see that it's hollow, an adaptation that allows the hair to be waterproof and insulative without being heavy. A cross section of a grizzly bear's hair, however, will appear as either solid or porous, with a few hollow pockets. The hybrid's hair is a weird combination of the two: a string of hollow pockets.

FEET

Many mammals walk on their toes (like dogs or cats) or hooves (like deer and horses). But bears walk on the flats of their feet. Other flat-footed walkers: musky carnivores (like wolverines and skunks), rodents (like mice and rats), marsupials (like opossums and kangaroos), red pandas, rabbits, raccoons, hedgehogs, and primates (like monkeys and humans).

The soles of the grolar's feet are only partially covered in hair, another blended feature. Polar bear feet are covered in hair to insulate them from the ice, whereas brown bears have hairless soles and clearly visible toes.

NECK

Grolars have longer necks more typical of polar bears, but also display small shoulder humps reminiscent of brown bears. A polar bear's longer neck helps it maneuver underwater, while a grizzly's hump holds fat to get it through

THROUGH DARWIN'S EYES

In *The Origin of Species*, Darwin waxed observational about the origin of whales, drawing a parallel between the mighty filter feeders and a bear he once saw swimming around with its mouth open, trying to catch bugs.

"I can see no difficulty in a race of bears being rendered, by natural selection, more aquatic in their structure and habits, with larger and larger mouths, till a creature was produced as monstrous as a whale."

Though he didn't have the whale's ancestry quite right, he wasn't entirely off base, at least from the bear's perspective. Technically, researchers have categorized the polar bear as a marine mammal, because it spends more time in the water than it does out. And as the arctic ice melts away, this is becoming increasingly true. Maybe we'll see the polar bear turn into a whale yet.

the winter. Looks like the grolar is hooked up with the best of both worlds.

LESSON OF THE POLAR BEAR: Don't believe everything you see. Don't believe everything you read. Species are dead; long live the hybrids.

TERMS DEFINED: Gene flow, species, hybrid

BLUE WHALE
(Balaenoptera musculus)

A HUGE PROBLEM FOR DARWIN?

Whales seem like something a confused kid made up. They evolved from land mammals that moved back into the sea, and then one of them (the blue whale) became the largest animal ever to live on Earth. Ever. Already an unlikely story. Now recent research shows that baleen whales may have been hybridizing across "species" for centuries now, further throwing shade on Darwin's Tree of Life. Are blue whales a bane or boon to Darwin's natural selection?

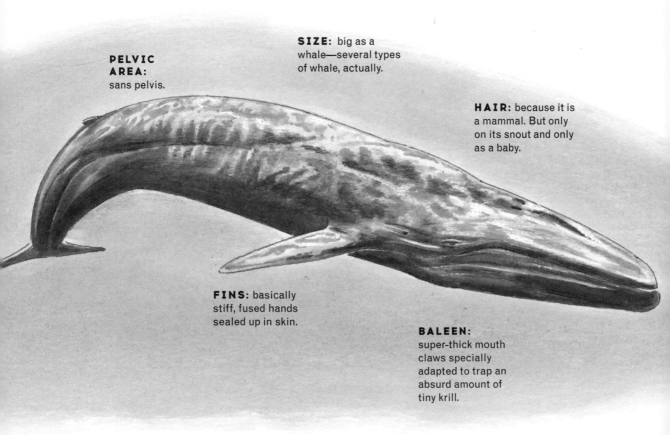

PELVIC AREA: sans pelvis.

SIZE: big as a whale—several types of whale, actually.

HAIR: because it is a mammal. But only on its snout and only as a baby.

FINS: basically stiff, fused hands sealed up in skin.

BALEEN: super-thick mouth claws specially adapted to trap an absurd amount of tiny krill.

WHEN IN (PRE)HISTORY/
WHERE ON THE RIVER

65 MILLION YEARS AGO

...diverged from the last mammal ancestor they shared with humans and most placental mammals.

56 MILLION YEARS

...early mammals went on evolving into larger land mammals with crude claws that lived by the water's edge. Groups of these animals diverged, and one eventually led to artiodactyls—the group of even-toed hooved animals that includes camels, giraffes, bison, cows, deer, elk, and moose. The other side would eventually lead to the hippopotamus and modern cetaceans. Yes, you heard right: cetaceans evolved from land mammals. Science has recently come up with a collective classification for the larger group of proto-hooved mammals that gave way to both the above groups: cetaceans + artiodactyls = cetartiodactyls. Blow your friends' minds by explaining that one over brunch.

53 MILLION YEARS AGO

...in what is now Pakistan, a long-nosed mammal about the size of a wolf lived by the water's edge, probably wading in to catch prehistoric fish and reptiles. One died and left a fossilized skull, whose ear bones are connected to the skull—most animals' ear bones float inside their soft tissue. The only modern animals with ear bones like that? Whales and dolphins.

44-49 MILLION YEARS AGO

...this mammal lineage had evolved mammals with an even more hydrodynamic face, more paddle-like feet, and a thicker muscular tail, all great for hunting in the water. Their nostrils migrated upward to the top of their heads, helping them stay submerged longer, and their hind legs became first pushed backward out of the way, then shrank and shrank until they were comically small, to be honest. Because many of these animals died in water where there is no water today, we have fossils and even complete skeletons for animals in the midst of this transition from terrestrial lifestyle back into the seas.

36 MILLION YEARS AGO

...cetaceans made their big split between those that would become toothed whales and baleen whales, called odontocetes and mysticetes, respectively.

28 MILLION YEARS AGO

...right whales and bowhead whales split from the other baleens.

SCALE
100 feet (30 meters) long, 200 tons (180 metric tons)

FAM-O-METER
70%?

A rough guess as to human overlap with whales, dolphins, and porpoises (together called cetaceans) comes via a partial genome comparison with dolphins and other animals. Baleen whales are likely to share even fewer genes with humans, as their bodies and lifestyles differ from humans' more than dolphins'.

Beyond that, scientists assumed that most baleen whales basically started to diverge any time after 10 million years ago, each filling different niches across oceanic territories of hundreds of thousands of miles. Today, there are about 10 recorded species of baleen whale, but these delineations are murky. A recent 2018 study conducted by German and Swedish researchers found that blue whales might have been mating and making hybrids with several other "species" throughout much of their evolutionary history. But even these findings were questionable: whales are extremely hard to ID in the wild and even harder to get DNA samples from, and since you can't keep whales in a lab most genome studies to date pull DNA from a single animal. But if the data is accurate, maybe great whales are another animal where hybridization is the norm and not the exception.

It makes logical sense, though. These are oceanic animals, the 2018 study points out: what's to stop them interbreeding if there are no geographic barriers keeping populations apart? Geographic areas mean zilch to whales like these, who migrate farther than any other mammal: in 2015, a female western gray whale was tracked traveling 13,988 miles (22,511 kilometers) in 172 days, breaking a humpback's previous migration-distance world record: 10,190 miles (16,400 kilometers).

It's also notable that baleen whales, or "great whales" as they were known in the olden days, were very nearly hunted to extinction before going on the endangered species list in 1978. Diminished population means fewer mating options. Back then, taking what you can get may have become a way of life. Now that the whales are saved, is there a benefit to continuing to fraternize? Genetic variation, of course. Options.

The more genomic information we collect, the more the scientific community seems to be adopting the idea that the "tree" of life might look more like a "web" or "network." (Here's where my river metaphor comes in again. That "gene flow" thing really applies, right?) Just like a real river, branches on the river of life sometimes reconverge. And what happens then? The genes in the river itself run deeper and wider and are less likely to dry up.

BEASTLY BREAKDOWN

SIZE

The second half of a blue whale's Latin name is a little joke: it comes from the word for muscle, but it's also the second name of the house mouse, *Mus musculus*. This may have been a little joke provided us by Linnaeus, Latin namer of animals, because the mouse is very small and the blue whale is the largest animal that has ever lived. As far as we know. Blue whales are larger in mass (and just barely about on par in length) with the largest dinosaur we've found yet.

HAIR/SENSES

Whales are mammals, and as such, they should have hair. Yet their aquatic lifestyle has long since made them bald.

Baby whales have hairs on their snouts when they're born, however, which spring from oversized hair follicles. The bumps remain even after the hairs fall out and the baby whale grows up; they're extremely sensitive to touch and can sense changes in water current, just like whiskers on a mouse do in the air.

ROSTRUM (SNOUT)

If you think a horse has a long face, get a load of this whale. Fossils show that the blue whale's land-walking ancestors had long faces, too, which may have been useful for hunting animals both onshore and in the water, like a crocodile.

Modern-day baleen whales fall into a few different categories in terms of how they feed, and their differently shaped rostrums may offer a glimpse into the evolution of feeding techniques. The earliest version of feeding may have been skim-feeding, like the right and bowhead whales do today, with their thick heads and deeply angled jaws. The next step may have been suction feeding, as the gray whale does today: instead of gulping water, the gray whale slurps silt off the sea floor, then filters out the sand, leaving behind plankton and crustaceans. Finally, blue and fin whales do full-on filter feeding, in which their flatter, streamlined bodies and wide mouths allow them to get the maximum bang for their buck when they open their jaws to scoop their food along with almost their own bodyweight in water. It pushes the water out again using its massive tongue, which is—you guessed it—the largest tongue on Earth.

Meanwhile the blue whale's close relative the humpback has perhaps the most high-tech method of expelling water from its baleen: rather than its tongue, it uses a massive, muscular throat pouch both to gather water and push it out again. Even though the humpback is not the largest animal Earth, its pouch is the largest single discrete biomechanism on Earth.

BALEEN

The baleen inside a whale's mouth might look like hair, but—well, technically it is made out keratin, the same stuff as human hair and fingernails. But the bristles are stiff and calcified enough to withstand sieving 4 tons (3.6 metric tons) of krill every day.

The exact origin of this unusual adaptation is unclear. But a 2018 study comparing great whale genomes to one another concluded that they likely moved from suction feeding to filter feeding around the same time that they began to diversify from one another and gain body mass to grow to their current stages of giantness.

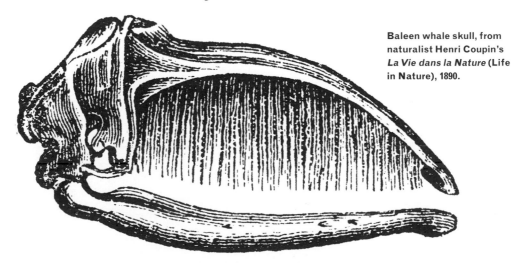

Baleen whale skull, from naturalist Henri Coupin's *La Vie dans la Nature* (Life in Nature), 1890.

BLOWHOLE

A whale's rostrum might look like its nose, but it's not, because to be called a nose, it requires nostrils. A whale's nostrils migrated back in the skull as it became more aquatic, allowing them to take breaths of air while remaining submerged.

IMMUNE RESPONSE

Baleen whales have excellent immune systems, harboring almost no major whale diseases in recorded science. Their newly discovered hybridization lifestyle may have something to do with this: the wider the variety of genes to choose from, the more adaptable the animal.

LOW-OXYGEN TOLERANCE

Part of the trick to a whale's deep dives is being able to withstand the exertion and pressure while holding their breath. It's no surprise, then, that they have genetic markers associated with tolerance of low-oxygen environments. These same markers show up in naked mole rats and other animals that live underground. They code for more robust pathways of oxygen delivery to the body's cells and even give the animal some control over how quickly it absorbs the oxygen it has already in its lungs.

FINS

Take an X-ray of a whale's front fins and you'll see a skeletal structure that looks a lot like your own hand—because it kind of is. Land-walking proto-whales had toes, ankles, knees, the whole bit. As they gradually made the water their frequent home, the fingers receded and were covered with a sturdy, hydrodynamic skin. The internal bones themselves have essentially become vestigial. But because the whale still uses the fins themselves to help it steer, the original components remain, even though their function is simplified.

Evolutionary "Speed" and Molecular Clocks: By comparing the changes in genome with fossil records, scientists are beginning to flesh out a new field of study that examines the timing with which genetic changes occurred. This helps them piece together two things: an overall relative timeline of the development of life on Earth, and an understanding of which animals may have evolved faster or slower than others—that is, which animals seem to have undergone the most genetic and physical changes within a set time frame.

TAIL FLUKES

Meanwhile, whales' hind limbs just up and shrank away, and their long bodies and tails became one long swimming powerhouse, undulating up and down, as compared to the side-to-side swing of a fish's tail.

PELVIC AREA

Whales no longer have legs, so they no longer have full pelvises—that bone saddle that your legs socket into. But they do have "pelvic bones," two small curved bones in their midsections, unconnected to the rest of their skeleton. For decades scientists thought whale pelvic bones were merely left over. Recent research suggests they might aid in certain mating behaviors.

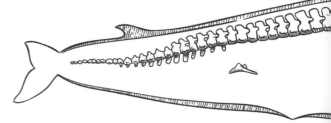

Whales do not, however, have a so-called "penis bone," which is a common misconception. Also known as a baculum, it's named after the Greek god Bacchus, notorious for his zeal for frequent and consensual sexual activity. Many other mammals have one, including cats, dogs, rodents, most primates, even. But whales don't, nor do their hippo cousins, elephants, nor any other ungulates. If you hear someone talking about a whale's "penis bone," they probably mean "pelvic bone." To be fair, they look alike.

STOMACH

Whales and dolphins have multichambered stomachs, and we know from a combination of fossil and genetic studies that they probably got them and kept them since before their lineage split with their hooved relatives. Modern cows and camels ruminate on their food—usually plants—moving it in cycles through their stomachs to eke out every bit of nutrition. Some even hock their food back up into their mouths to chew as cud before swallowing again. But whales eat fish and crustaceans, and they don't chew anything, much less cud. So why the stomachs?

The whale's distant cousin the camel is also not a true ruminant in that it doesn't chew cud. But its stomachs, not just its fatty humps, hold the secret to their famous between-meal survival. They've evolved to be able to take in relatively more food or water in one sitting than most other animals on Earth, more than their

bodies can digest all at once. So they keep the food in various stomachs in various stages of digestion, which in turn gives them more to draw from later, if food is scarce—like keeping leftovers in your belly instead of the fridge.

Blue whales do this too. Krill, which are like tiny shrimp, show up in large schools, so whales scoop up as many of them as fast as they can. This daily dose of 40 million little invertebrates will then pass them through the whale's system at a measured pace. They've retained their ancestral stomachs. It seems that evolution has a rule: if you use it, you won't lose it.

But several studies show that they *have* lost a special metabolizing gene since the split with their plant-eating cousins. The gene encourages

Convergent Evolution: Coincidental Co-incidence

Laypeople used to think a whale was a giant fish, and for obvious reason. It lives completely submerged for most of its life and sports fins and a fanned tail. When unrelated animals evolve to have traits that are similar in appearance because they serve a similar function, that's called convergent evolution. Think wings on bats, birds, and flying insects. No close relationship among those three, but all took to the skies, and all have wings.

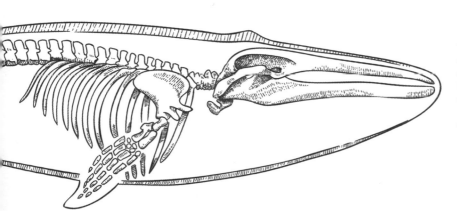

Think of the whale's skeleton as that of an early land mammal that has been Photoshopped, then run through several million-years'-worth of face-tuning app filters.

THROUGH DARWIN'S EYES

Evolutionary biologists love moments of change at the water's edge: the sea sludge that learns to photosynthesize; the fish that learns to walk on land. Cetaceans (whales and dolphins) are a great model to look for telltale signs of evolution because they did it "backward"; they'd made it out of the water, but then they went back in. For land mammals like us, it's almost hard to swallow.

In 1859, when Darwin dropped *The Origin of Species*, the general public still believed whales to be enormous fish. Scientists knew better by then, but the average person had no reason to see a whale from the inside, unless they were a whaler. Darwin's comment about the bear swimming around with its mouth open only hurt his case. The newspaper clowned Darwin so hard on this particular image of a bear swimming about that

he eventually removed the reference from later editions of his book. But we now know that he had nailed the general idea. In fact, the truth—that whales share ancestors with hooved herbivores—is even stranger.

Fast-forward to 2018, the day the blue whale hybrid study drops. Like in Darwin's time, the press has a field day. The internet exploded with headlines reading "A New Study on Whales Suggests Darwin Didn't Quite Get It Right" and "Why Darwinism Is False." True, Darwin had pinned much of his working theory on natural selection as the driver for change in animals, but he hadn't limited it to that. Darwin was never satisfied that his model was complete. A less sexy but more accurate headline might be: "Darwin Limited by Technology Available to Him at the Time!"

Comparative Anatomy: Seeing Is Believing

Being able to compare genomes is a game changer when it comes to understanding evolution. But there's still something pretty astounding about some good old-fashioned comparative anatomy: a side-by-side examination of the analogous parts of an animal's body.

NOTE: analogous body parts and genes are not the same thing as homologous body parts and genes (see definition, page 147).

a special enzyme production in the pancreas, an enzyme that helps the animal break down challenging material like plant matter and potential toxins, and it shows up in a variety of ruminating animals, including cows, camels, and even some monkeys, bats, and one unfortunate bird (the hoatzin). But cetaceans, they didn't use it, so they lose...d it.

Around this same time, it seems that whale genomes started making other changes as well. Gene sequences related to fat storage and energy production from fat seem to have evolved after they split from the majority of their land-animal relatives. Hmm.

BLUBBER

Blubber is a special layer of insulating fat found in sea mammals like whales, seals, and walruses. Whale oil, which was extracted from blubber, was a valuable fuel as early as the 1820s. It was the driving factor behind the whaling industry, which extended through to 1978 and almost wiped out most of the world's large whale species, collectively called "great whales," which included the larger baleen whales and the large and toothy sperm whale.

But what didn't kill them made them stronger. The recent hybrid findings tell a whale of a tale (I couldn't help myself). While the great whales were being hunted to near extinction, they were getting creative in their approach to keep their gene pools from drying up. Sounds like natural selection to me. And species originating species.

LESSON OF THE BLUE WHALE: Hybrids again, eh…? Maybe this is more of a thing than we thought. Also, whales used to live on land, and that is crazy.

TERMS DEFINED: Comparative Anatomy, Convergent Evolution, evolutionary speed, molecular clocks

KNOWLEDGE OF THE LEVIATHAN

For thousands of years, the world's only knowledge of whale anatomy was carried by whalers alone, from the Makah tribe of the American Northwest to medieval seafarers in Europe and Asia. Whalers kept detailed record of whales' movements and dissected them seaside or deckside each time they made a kill, building knowledge of whale anatomy. The author Herman Melville famously drew from his experiences aboard whaling ships for his 1851 novel *Moby Dick*. Though the book was fiction, it included an entire chapter titled "Cetology"—the science of whales—which has gone down as one of the first major publications on actual whale science.

Whale-Fishing, from the Cosmographie Universelle, 1574

TRINIDADIAN GUPPY

(Poecilia reticulata)

TEACHER'S PET

Now that we've considered the origin of "species," let us examine what Darwin meant by "means of natural selection."

The term *guppy* refers to a single species of fish, but their types and characteristics hit every color of the rainbow, with varying levels of fanciness. They're prolific enough to serve as classroom pets that conveniently complete their reproductive life cycle in the span of a three-week biology unit. The most famous, though, in scientific circles anyway, is the Trinidadian guppy, once central to one of the most groundbreaking studies in the history of evolutionary biology.

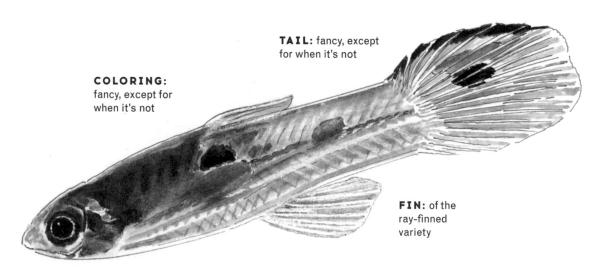

TAIL: fancy, except for when it's not

COLORING: fancy, except for when it's not

FIN: of the ray-finned variety

TEETH: not very big or intimidating, hence the coloring/tail situation

WHEN IN (PRE)HISTORY/ WHERE ON THE RIVER

Guppies are a fast-evolving member of the largest and most recently evolved fish group, called teleosts, or sometimes "true fish" or "bony fish." This is also the most diverse group of fish and, in fact, the most diverse group of vertebrates, with more varieties of fish than in any group of reptiles, amphibians, or mammals. Teleosts include everything from flatfish like halibut to game fish like marlin, angel fish, cat fish, and even deep-sea angler fish, with their terrifying maws and light-up lures.

The guppy broke away from the rest of teleosts about 25 million years ago. It was first discovered in the Caribbean by Western scientists in the mid 1800s, around the same time that Darwin was crossing oceans on the *Beagle*.

BEASTLY BREAKDOWN

COLORING

In 1975, an American researcher named John Endler was doing field work in Trinidad when he noticed that guppies of the same breeding species look very different from one another in different streams. In some streams, the males were brightly colored, with orange and blue spots and big flashy tails. In others, guppies of the same species as the colorful ones had drabber coloring—grayish brown with just small dabs of color—and smaller dorsal fins and tails. Even guppies in the same stream vary from bright to blah, depending on the part of the stream they live in. This, Endler thought, was curious.

Endler had been studying evolution a long time by then, and he knew that coloring often relates to sexual selection: typically males showing off part of their anatomy to attract females. The flashier males get to mate and reproduce. But usually, in a case like this, the male's attribute keeps getting bigger and/or brighter, like peacocks or the massively antlered Irish elk (aka giant deer, now extinct). It seems odd that *some* males in the same relative population should be getting fancier while others get actively less so.

Usually when members of a species get actively less fancy, a little more like their

---

SCALE
Males: 0.6–1.4 inches (1.5–3.5 centimeters) long, females: 1.2–2.4 inches (3–6 centimeters) long

FAM-O-METER
69%?

The genetic overlap between human and Trinidadian guppy has yet to be calculated. But the human and the zebrafish share a remarkable amount of genetic information, many millennia after separating on the river of life. Overlap with the guppy might be similar.

WHAT'S THE DIFFERENCE?

GENETICS is the study of genes (aka sections of DNA, aka groupings of DNA/RNA base pairs).

WHOLE GENOMICS is the study of an organism's entire collection of genes (aka all of their DNA, or rather the template to all their DNA).

Science has studied *genes* longer than we've been able to sequence whole genomes. That is, scientists coined the term gene before we really understood what DNA actually is. *Gene* is, therefore, a catch-all term that refers to "the thing that holds the information that I get from my mother or father," or "the thing responsible for a trait like brown eyes or a curlable tongue." We understood by the 1940s that "genes" showed up in the chromosomes (rod- or ring-shaped threadlike structures inside animal eggs and sperm), but the word was a placeholder for the mechanism of inheritance itself, before we figured out what that was. It was DNA/RNA. DNA keeps the information; RNA puts it to work.

Once we figured out that the mechanism was DNA/RNA, it may have made more sense to scrap the term *gene* and dig into DNA with fresh eyes. But we kept the term *gene* and applied it to chunks of DNA that clearly did something, chunks that we understood. Chunks of RNA that were clearly associated with making something tangible like a protein or an enzyme we called "coding" genes. Chunks of DNA that we couldn't match to a job we called "non-coding" or "junk" DNA. Our terms, mind you, our definitions of *job* and *code* and *tangible*.

---

surroundings, it's because of predator selection. The better the camo, the more likely you are to survive and reproduce. But in a population like that, eventually everyone blends into the background.

This population seemed to be working both ends of the spectrum. But also, neither end was taking it to the extreme.

After months of drawing detailed maps of streams and pool location, photographing and recording guppy size and coloring, Endler began to see a pattern emerge: Areas of the stream where guppies were duller were *also* home to other, predatory fish. Areas where guppies were flashier had fewer or no predators. Likewise, the females of the species preferred mating with colorful males, but only in areas where their offspring weren't at risk of being noticed and eaten. The guppy mystery had a two-pronged solution: the same species were under two simultaneous selective pressures. But the best was yet to come.

Endler's colleague, who had been assisting with the research, used the opportunity to turn field observational research into experimental research. Long story short: he built huge tanks in a lab setting and populated them with guppies, taking advantage of the guppies' short life cycle to artificially control survival factors *except* predators and see the effect over time. He wanted to see if the findings of the field study could be reproduced in the lab. And it worked:

over several generations, the guppies began to diverge in coloring, just as they did in the wild. It was controlled evolution in real time.

In the tradition of Gregor Mendel and his pea experiment, these researchers showed that evolution can stand up to the same scrutiny as other experimental sciences. (Although, if you ask any farmer or pet breeder, they'd be right to tell you that they experiment with evolution every day of their lives.) But until this point, many "hard" scientists had considered evolutionary biology to be a "soft" science, like anthropology or sociology, based on observation rather than verifiable, numerical data. The guppy experiment catalyzed a new movement in evolutionary biology, and soon researchers were controlling the evolution of bacteria, mice, anole lizards, zebrafish, frogs, mice, and rats. One long(est)-term study using *E. coli* bacteria started in 1988 and is still continuing. It has tracked changes in more than 68,000 generations of bacteria, the equivalent of one million years of human evolution.

Experimental evolutionary biology eventually led to the hard, verifiable field of genetics, then whole genomics. Genomes give insight into what *did* happen, though they rarely come with a roadmap of how. Now that we can align genomic data with experimental controls in the lab, the roadmap is slowly coming together. Though it will prove to be far more complex than the streams and pools of the Trinidadian guppy.

LESSON OF THE GUPPY:
Evolution can not only be observed, it can also be tested and proven in the lab.

TERMS DEFINED: Sexual selection, predator selection, genetics, genome

THROUGH DARWIN'S EYES

Maybe you've heard the story of Darwin's finches. Observing a flock in the Galapagos Islands, he noticed that finches with certain beak shapes only fed on certain types of food. Finches with large, curved beaks fed on large nuts that required a powerful bite. Those with small, pointed beaks fed on small insects. A beak for every food and every food in its beak.

The observation was, as the story goes, Darwin's eureka moment, the very basis for his term *natural selection*. Like predator selection or sexual selection, scientists have since refined the finch beaks as literal textbook examples of niche selection, niche adaptation, or adaptive radiation.

It seems likely that if Darwin could have proven his hypothesis by taking the finches home, feeding and breeding them in a controlled environment, and speeding up their life cycles to observe changes to their physical bodies over many generations to see if their beaks changed as they found their niches and subsequently survived to procreate, he would have.

It's worth noting that a pair of biologists did in fact follow up on Darwin's finch hypothesis. Married couple Rosemary and Peter Grant lived in the Galapagos for almost 30 years, painstakingly recording finch feeding behavior and beak measurements. Together, patiently, they were able to prove with hard data the theory that birthed the theory that revolutionized our understanding of life on Earth.

SCALE
Adult male
giraffes range
from 16 to 20 feet
tall (14.9–16
meters)

FAM-O-METER

79%?

Until the complete giraffe and
human genomes are compared
side by side, we can't know the
exact overlap between us. This
estimate comes from our
overlap with cows, minus a
little relatedness.

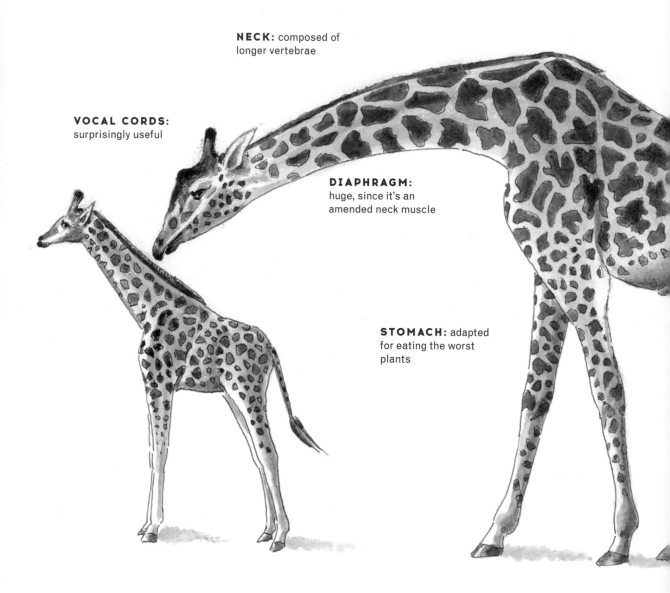

NECK: composed of
longer vertebrae

VOCAL CORDS:
surprisingly useful

DIAPHRAGM:
huge, since it's an
amended neck muscle

STOMACH: adapted
for eating the worst
plants

MASAI GIRAFFE

(Giraffa camelopardalis tippelskirchi)

IT'S NOT ALL IN THE NECK

The giraffe's long neck is a feature so obvious that it's been the subject of speculation for ages. Masai tribes of central Africa passed down observations about the giraffe centuries before European scholars began using it as the perfect target for philosophical small-*t* theorization. But it took genomics to finally tell the story of how the giraffe got its neck.

WHEN IN (PRE)HISTORY/ WHERE ON THE RIVER

Four species of giraffe live throughout sub-Saharan and southern Africa. Today they're largely geographically separated, and they haven't interbred for around 2 million years.

65 MILLION YEARS AGO

CENOZOIC

...20–25 million years ago, the okapi/giraffe lineage split off from the larger group of hooved animals that eventually gave way to deer, antelope, and moose.

MASAI GIRAFFE

...27.6 million years ago, proto-okapi/giraffes split from the cattle lineage that eventually became cows, bison, oxen, and water buffalo.

...11.5 million years ago, giraffes as a group split off from their closest extant relative, the okapi.

BEASTLY BREAKDOWN

COLORING

For centuries, scientists assumed giraffes all over Africa were all the same species. Sure, they were geographically isolated from each other and had some differences in their appearance, but a giraffe is a giraffe. One of a kind. Or…four of a kind? In 2016, an international group of researchers examined the DNA of 190 individual giraffes from a variety of populations around Africa. Their findings showed four different lineages of giraffe, "as genetically different as the brown bear from the polar bear," said one researcher. (This comment, though, may be more complicated than it sounds, see page 29.)

What was once simply the giraffe became the southern giraffe, Masai giraffe, reticulated giraffe, and northern giraffe—four species where once there was one.

As we learned from the platypus, animal identification relies on whatever observational tools are available: noting an animal's outside when you could see it and insides when you could catch it and preserve it. Back in the day, distinctly different coloring might have earned each group of giraffe its own species name. There are arguably nine to eleven separate subspecies of giraffe, so called because the differences among them are largely superficial, having to do with the pattern of their coat.

But the age of the genome changes the rules yet again. As our identification tools increase, so too does the number of categories increase. In the age of genome sequencing, when differences are no longer skin-deep, the difference between calling out a new species versus a new subspecies of giraffe is up for a lot of debate. They haven't interbred for at least two million years. They

likely could, with viable offspring, but they don't.

NECK

If you had to build a longer neck in a hurry, you might think to add extra neck bones: more vertebrae for more length, right? But that's not the case. Humans have seven vertebrae in the neck. And giraffes have seven neck vertebrae, too. In fact, all mammals have seven neck vertebrae, except sloths and manatees. In giraffes alone, though, they're extra elongated.

One way to understand how they got that way would be to look at the fossilized neck bones of ancient proto-okapi and proto-giraffes. In 2014, a New York anatomist and her colleagues closely examined the second and third vertebrae on 71 giraffes and proto-giraffes from 11 species. Some were from fossilized specimens from as many as 16 million years ago. And eureka: even the old ones had neck vertebrae with a relatively longer length-to-width ratio. It's almost like we knew it when we categorized them as "giraffe." The neck is part of what makes a giraffe a giraffe.

So the question isn't "How did the giraffe get a long neck?" because the giraffe has always had a long neck. But we might then ask, "How did its genetically different predecessors get theirs?"

The next logical question would be "How might a long neck benefit a proto-giraffe?" And the obvious answer might seem to be: leaves. Food. When food is scarce, the taller animal can reach more leaves.

But in reality, when leaves are scarce, modern-day giraffes actually forage for plants elsewhere. In the underbrush, for instance, in which case their long necks actually impede them. So a long neck must have helped them in some other way than "high-up leaf getting."

Recently, a new theory has emerged: perhaps the long neck became a way of attracting mates and defending territory against rivals, a product not of niche selection but rather sexual selection. Indeed, there's no giraffe version of Darwin's finches, with shorter and intermediate giraffes each feasting on different foliage.

So rather than a finch's beak, adapted for eating, a giraffe's long neck is an adaptation for the reproduction part of the "getting ahead." Its neck is more comparable to a moose's massive set of antlers or a peacock's dazzling array of tail feathers. Indeed, modern-day male giraffes do engage in neck-based competition to win their mates, slapping one another with their necks like two much clumsier stags would with their antlers. "Necking," as the activity is unfortunately named, requires not just a long neck but a strong one, too—which does correlate with the lengthening of vertebrae. The giraffe found

success with a longer, stronger, stiffer neck, supported by hefty bones. Compare this to the delicate and flexible neck of a swan, with 25 neck vertebrae. Some necks are just made for battle and some aren't.

SKELETON

Other parts of the giraffe's skeleton look different as well: its legs, for instance, because just imagine how silly if they weren't. And the nuchal ligament, which holds the head upright in adult animals, is enlarged and strengthened, in order to sustain the weight of the long neck and head.

The comparative giraffe genome came out of a massive 2016 comparative study conducted by researchers from Tanzania and the US. They found that the majority of differences between the okapi and giraffe genome(s) have little to do with the appearance of the giraffe but make crucial changes having to do with the neck change. Is there a neck gene? There is not. But a neck isn't simply a "neck"; it's bones, muscles,

POSITIVE SELECTION

When a trait seems to have gotten more play over time, like when a trait shows up in a population more and more over time, that's called positive selection. The idea here is that if a trait shows up more and more over time, it might actually have something to do with helping the animal survive or mate. Either way, something about the trait is helping the trait (or, as we'll later learn, gene or collection of genes) appear more often over time.

Because of that word *positive*, it might sound like the trait is actually making the animal's life better in some way. And, well, if

it's helping it survive or mate, it is. But the term has less to do with quality of life and more to do with "more of this thing is happening." In the case of the Trinidadian guppy, it's complicated: fancy tail feathers might be positively selected in a pool with no predators, but if a predator suddenly appears, the same trait loses that positive upswing real quick. In the case of the giraffe, it's also complicated. Longer vertebrae were "positively selected." But the more we learn about genetics, the more we see that such upswings are driven by more than just who's eating where and mating with whom.

blood vessels, blood flow, and electrical nerve messages to the brain and back again. It's genes that keep the embryo of the giraffe from stunting by multiplying differently than a shorter animal might, assigning the ratio of muscle to cell to bone to connective tissue cells in unusual amounts. It keeps the metabolism high, which means changing a cascade of functions all the way down to the organelles inside cell nuclei. It's many adaptations that have to work together for a fully functional giraffe.

The study calls these MSA genes, which just stands for "multiple signs of adaptation." Remarkably, nearly half of these genes are involved in controlling developmental pattern formation and differentiation—the "what kind of cells and messages go where when" stuff. These types of genes are largely shared throughout the entire vertebrate kingdom, as you'll see as you continue through the book. And they each have their own name, given to them by the researchers who discovered them at different junctures. Some of them are about as creative as MSA gene (fibroblast growth factor or FGF pathway genes, transcription factors E2F4, E2F5, ETS2, TGFB1, folate receptor 1 or FOLR1), while others have names that sound like video game characters: Notch, Homeobox, and Sonic Hedgehog. Literally. All of the above genes exist in every limbed animal—humans, too (see "Jungle Fowl," page 193).

Thanks to quiet adaptations to all of these gene groups, the rest of a giraffe's body managed to make the many adaptations necessary to give it a long neck and still survive.

HEART

Because of its long neck, a giraffe's heart has to work harder than most animals' to pump blood against gravity and maintain blood pressure homeostasis. It has a relatively bigger heart than most animals of comparable size, and the muscles in the walls of its heart are extra thick and strong.

BLOOD VESSELS

The blood vessel walls in the lower extremities are greatly thickened to withstand the increased hydrostatic pressure, and the venous and arterial systems are uniquely adapted to dampen the potentially catastrophic changes in blood pressure when a giraffe quickly lowers its head to drink water.

DIAPHRAGM

The giraffe has the largest and strongest diaphragm of any animal. It's not because it requires a different type of air up there, but because the diaphragm is actually an oddly adapted neck muscle that migrated down into the chest cavity in mammals. Bigger neck, bigger diaphragm by default.

STOMACH

Mitochondrial metabolism and volatile fatty acids transport genes are different between the giraffe and the okapi, which may be related to its slightly different top-of-tree diet. Rather than finding a simple solution by eating up top, giraffes may have in fact opened themselves up to eating new and potentially toxic plants. Luckily, giraffes are ruminants, meaning they process their food really slowly and through multiple stomachs, to reduce the chance of being poisoned by a new and unfamiliar toxin.

VOCAL CORDS

Researchers have recorded giraffes making what seem to be communicative hisses and grunts. But they long assumed that even with a horse-like larynx (voice box), giraffes' 3-foot-long (1-meter) trachea (windpipe) would be too long to produce anything beyond infrasonic (ultralow) sounds, below the range of human hearing. But then, in 2007, a zookeeper in Austria swore he heard his giraffes *humming* late at night when zoo visitors weren't around to hear. Researchers spent the next eight years

THROUGH DARWIN'S EYES

It's interesting that history has remembered the "top-of-the-tree leaf-getting niche" as the advantage Darwin attributed to a giraffe's long neck. In his most landmark work *The Origin of Species*, Darwin sets up the telling of his giraffe argument through the story of debate held among him, his colleague A. R. Wallace, and several other public intellectuals at the time. In it, the group discussed the many advantages a long neck might afford a giraffe, including "necking":

Assuredly the being able to reach, at each stage of increased size, to a supply of food, left untouched by the other hoofed quadrupeds of the country, would have been of some advantage to the nascent giraffe. Nor must we overlook the fact, that increased bulk would act as a protection against almost all beasts of prey excepting the lion; and against this animal, its tall neck,—and the taller the better,—would, as Mr. Chauncey Wright has remarked, serve as a watch-tower. It is from this cause, as Sir S. Baker remarks, that no animal is more difficult to stalk than the giraffe. This animal also uses its long neck as a means of offence or defence by violently swinging its head armed with stump-like horns. The preservation of each species can rarely be determined by any one advantage, but by the union of all, great and small.

recording over 930 giraffes at three different zoos and analyzing the vocalizations to find some proof of intentionality in the humming behavior, which meant looking for a pattern. Eventually a pattern did emerge, but the reason why the giraffes make this sound is still unclear. To look for clues, the researchers next want to compare the vocalizations with night-vision observations of the giraffes' physical behaviors.

But this is where having a genome to reference comes in handy. Where might giraffe genomes overlap with animals known to communicate vocally, like humpback whales or wolves? What about genetic overlap with non-nocturnal animals that have nocturnal behaviors? What else in nature happens at around 92 hertz and might be associated with genetic markers? What about genes implicated in aural processing? I did a quick dive and found that giraffes do show genes connected to aural processing, which isn't unusual: they share them with most other prey animals in their genetic neck of the woods. They don't however, share the FOX gene associated with speech (more on this later). So the jury's still out on the mystery of the humming giraffe. Researchers refer to unanswered questions like this as a "recommendation for further study."

LESSON OF THE GIRAFFE:
Even superficial traits are more than skin-deep. Visible traits are often linked with unseen traits in ways that only a genome can illuminate.

TERMS DEFINED: Positive selection

TEETH: long, strong, and one of the reasons it's been around so long.

LEGS: adapted to be better at standing and running than not.

HOOF = TOE: Yes. The hoof is one giant toenail on one giant toe.

HORSE
(Equus caballus)

BORN TO RUN

The horse's genetic and evolutionary history doesn't add anything new to our story thus far: hybridization is definitely more of a thing than we'd imagined, and certain gene groups code for certain traits that come in handy. For the horse, it's all about teeth, nerves, and legs that don't quit.

WHEN IN (PRE)HISTORY/ WHERE ON THE RIVER

Equus caballus includes all domesticated horse breeds, from skinny thoroughbred racehorses to stocky plow horses to tiny ponies, which belong to a single species. All surviving branches of the horse family tree are also members of this same genus, *Equus*, which now consists of only seven living species, including zebras, donkeys, and asses.

Horses have looked like horses for a long time. The oldest equine fossil dates back to around 55 million years ago and was found in North America. Most horse lineages later migrated to Asia, only returning when European colonizers brought them back on ships.

In 2006, a surprising finding comparing the genomes of horses and their near relatives found something unexpected: strings of repeated "junk" DNA suggested that horses' closest living relatives outside of rhinos and tapirs are bats, then dogs.

Horses' recent genetic changes hinge largely upon their long relationship with humans. The first human cave paintings in Europe feature horses.

But even the modern domestic horse genome shows a whole lot of interbreeding with wild horses. Here again, the boundaries between animal groups seem arbitrary to the animals in them.

THROUGH DARWIN'S EYES

The distension of the nostrils is not for the sake of scenting the source of danger, for when a horse smells carefully at any object and is not alarmed, he does not dilate his nostrils. Owing to the presence of a valve in the throat, a horse when panting does not breathe through his open mouth, but through his nostrils; and these consequently have become endowed with great powers of expansion.

Although he didn't know it at the time, Darwin's observation touched upon the findings of a 21st-century study that determined horses have an especially reactive sensory system, especially in their skin. Such sensitive skin reactions may have kept horses safer from biting insects and potential infections over the millennia.

Darwin on his trusty horse Tommy, 1870s.

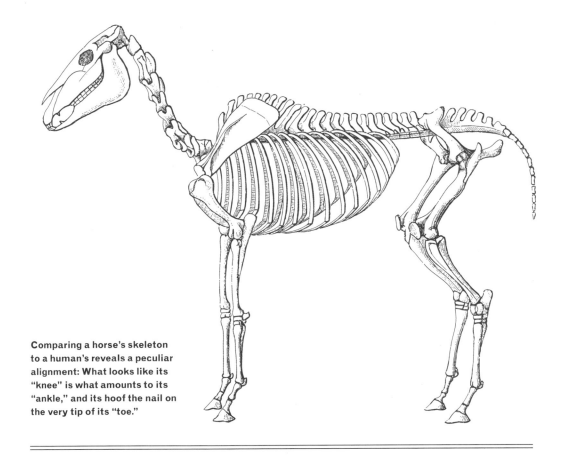

Comparing a horse's skeleton to a human's reveals a peculiar alignment: What looks like its "knee" is what amounts to its "ankle," and its hoof the nail on the very tip of its "toe."

BEASTLY BREAKDOWN

HOOVES (AKA TOES)

Horses, humans, and all other mammals share a common ancestor with five toes. The horse's hoof is actually a single middle toe, highly adapted, with a single massive toenail for a hoof.

TEETH

As horses adapted to eating tough grasses, their teeth became stronger and longer. But being long in the tooth actually helped the few extant equine species survive the major climate shift during two ice ages.

LEGS

Horses' legs are evolved to have incredible stamina, made possible by groups of ligaments that minimize the energy going into the leg to get it to function. This is why horses can sleep standing up. In a 2013 study, Korean researchers found that horses had more complicated genes coded for connective tissue than in other mammals their size.

LESSON OF THE HORSE: If it works, run with it.

PROBOSCIS: a mouth tube possibly evolved from something stabbier

EYES: actually really not meant for night vision

COLORING: depends on its environment

FEET: which smell/taste

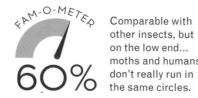

SCALE
Wingspan of about 2 inches (about 53 millimeters)

FAM-O-METER

60%

Comparable with other insects, but on the low end... moths and humans don't really run in the same circles.

PEPPERED MOTH

(Biston betularia)

WANT TO PAINT IT BLACK

Few things are black and white when it comes to evolution, but the peppered moth's story is about as close as you can get. Heck, it's been used as a classic example of environmental selection since movies only came in those colors. But thanks to new understanding of *how* genes change (aka what some "mutational events" look like in the genome), this dusty old tape is playing out in full color.

WHEN IN (PRE)HISTORY/ WHERE ON THE RIVER

I once heard a biologist explain the combined group of butterflies and moths (Lepidoptera) like this: All butterflies are technically moths, but the combined group wouldn't be complete without butterflies as part of it. There are 160,000 species of them in all.

Fossil evidence shows that modern-looking butterflies have been in North America for at least 34 million years.

In Lebanon, a 125-million-year-old caterpillar was fossilized in amber with a modern-looking spinneret, the body part that all butterflies and moths use to spin themselves into a chrysalis or cocoon, respectively.

The oldest known proto-moth fossil dates from at least 200 million years ago, identifiable by scaly wing impressions and a few other key traits. This calls into question the notion that Lepidoptera evolved from a common ancestor of caddis flies, possibly the most famous prehistoric insect next to dragonflies, which appear in the fossil record as recently as 250 million years ago.

BEASTLY BREAKDOWN

COLORING

The peppered moth is white flecked with black, hence its name. But which color is most common? That depends on its environment. And in this particular case study, the peppered moth's environment is the major city of London, England.

The peppered moth's typical color, off-white and mottled, blends perfectly with the bark of a birch tree, which are common in England. The black or carbonaria-colored moths also emerge, but have a harder time hiding out on white birch. But at a particular moment in time, they had their moment in—er—out of the sun.

London, 1850. The Industrial Revolution was in full coal-burning swing, coating everything in a thin layer of soot. The peppered moths' habitat became increasingly black, from building eaves to city-adjacent groves of trees, which, in turn, made the pale moths easily visible targets to predators such as birds and bats.

That's when the dark-type peppered moths started to take over. Over the next few years, the whole population turned dark from the inside out: the pigment of their actual wing scales, white in previous generations, turned black through increased melanin.

By 1953, a young British researcher noticed the two colorings and decided to capture, mark, release, and recapture moths of each color, in both the polluted inner city and less polluted countryside. His experiment was not particularly rigorous, but he correctly hypothesized that the darker moths "disappeared" more readily in the city and vice versa. London passed the Clean Air Act in 1956, and over the next 20 years, the researcher cataloged the area's peppered moths.

Moving through his collection today, you can watch the shift back to pale before your very eyes.

A half century would pass before the mechanism for the change would be determined. In 2016, Liverpool researchers attempted to re-create the original experiment under more controlled laboratory conditions. They compared genomes between the pale and dark specimens and isolated the ability to a small section of genome whose job it is to move around in the genome and "switch" specific other genes on and off.

These sections of DNA are called transposons, or "jumping genes." And they're kind of a big deal.

Transposons: aka transposable elements, TEs, or even jumping genes. This is a term so new in the field of genomics that the field hasn't even agreed on the universal term to stick with. In the early days of genomic science, DNA that didn't code for a specific protein or amino acid was labeled "junk" DNA. But around the time of the Liverpool moth study, researchers from throughout the field are finding that the so-called junk DNA has a job after all. Or rather, it has several jobs, all of which we have yet to discern. The secret lives of transposons, now that's a "recommendation for further study" that is rich with opportunity in the coming five or so years.

The Liverpool lab named their color transposon cortex, after the part of the brain that makes all the decisions.

This idea of genes making "decisions" is a slippery slope. They're obviously not sentient. But the idea that the genome has reserves of genetic information that can change during the lifetime of an animal is a new notion. Evidence for the flexible job of a transposon/jumping gene has only begun to emerge in the second half of this century.

POISON CONTROL

Several species of moth and butterfly have the unique ability to live off of plants that are toxic to other animals. In the case of the monarch, it lives and breeds on milkweed, which makes it taste bitter and unappetizing to potential predators. A 2015 study suggests that this ability is another case of coevolution, when two species affect adaptations in one another. The study's authors even go so far as to suggest that anything that eats plants has to have coevolved with those plants and to have evolved some kind of anti-toxicity system. (I'm taking this to mean that for humans, vegetables are not to be trusted.)

Likewise, viceroy butterflies are not to be trusted. They avoid milkweed and taste just fine, but they've coevolved to copy the patterns of monarchs in their region so as to avoid predation, too. The authors of the peppered moth study put out a sister paper finding that the same section of genome, cortex, helps mimic butterflies like the viceroy get their coloring right.

PROBOSCIS

In late 2017 a clog of Dutch geologists were trying to drill for ancient pollen embedded deep in sedimentary rock, when they stumbled upon a nearly 200-million-year-old fossilized proto-butterfly/moth. They easily recognized it as one of a large group that share a proboscis: a long tube with a mouth on one end, which serves as a sort of big straw for slurping nectar from deep inside flowers. The weird part is that common wisdom holds that plants didn't evolve flowers until about 100 million years ago, about 100 million years after the animal that created that fossil died.

It would seem to be a trait born from the coevolution of nectar eater and nectar maker, but that doesn't work if one pre-dates the latter.

THROUGH DARWIN'S EYES

Darwin was so into insects coevolving with plants that he wrote a whole book on orchids and their pollinators. He had observed that orchids and moths in particular had a beautiful symbiosis, a connection so strong that he once predicted the existence of a moth without ever having seen it, based solely on the unusually deep nectar well of an orchid. He wrote in a letter to a friend, "I have just received such a Box...with the astounding *Angraecum sesquipedalia*... with a nectary a foot long. Good Heavens what insect can suck it." Later, in his orchid/insect book, he wrote a full hypothesis as to what the sucker might have to look like...And sure enough, someone discovered that crazy-proboscised moth in 1903, 21 years after Darwin's death.

Coevolution: Another Co-incidence

Like other types of evolution, coevolution doesn't have intent or trajectory. The favorite flower of a butterfly didn't intentionally evolve to emphasize colors the butterfly could see. Nor did the butterfly intentionally evolve a fancy bonus sense for detecting extra-spectrum colors to find the best untouched flowers. By tiny fits and starts, small changes occurred in each, animal and plant, and those changes affected both for the better. As we'll see in the description of red coloring in finches, there may be some sort of causal relationship between those slow changes, but so far they remain a mystery to science.

While it's possible that the botanical fossil record just hasn't turned up any flowers yet—ever—it seems much more likely that it's the purpose of the proboscis, not the entire paleontological history of plants, that needs to be rethought here.

The new working theory has to do with the heating-up period that Earth was undergoing at the time. If water was scarcer, maybe an animal with a built-in face straw could get to water collected in nooks and crannies. If the mosquito is any indicator, maybe the proboscis had more of a stabby/poky/drainy function before flowers gave insects somewhere to stick it. Or, if we caution ourselves against the "why" and the "what for," maybe the face straw correlated with some other genetic mutation that helped some proto-moth in the time before the Triassic (200 million years ago).

EYES

Like many insects, moths have compound eyes, which are what they sound like: a collection of

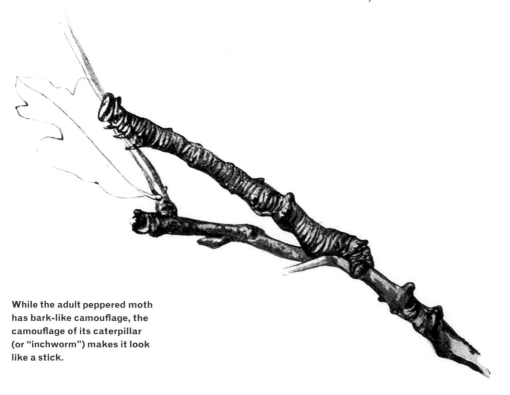

While the adult peppered moth has bark-like camouflage, the camouflage of its caterpillar (or "inchworm") makes it look like a stick.

many different lenses, pressed together into one. Each lens is not as complex as the camera-like eye of say a cat or a human, so the compound eye doesn't get a high-def picture. Rather it gets a more three-dimensional understanding of its surroundings as it navigates through the air.

At the same time, the collection of many lenses means that it can do other stuff, like see extra-spectrum light, such as ultraviolet and infrared. Pollinators like butterflies and flowering plants coevolved, meaning that both organisms adapted and evolved in ways that benefited both themselves and each other. It's as though flowers and their pollinators exchange secret messages—"for pollinator eyes only."

Moths, however, do have some problems with their vision. As nocturnal animals, they have evolved to orient themselves via a distant light source like the moon. But with artificial light mucking up the works, they start to fly toward the light and bump right into it.

CHEMICAL SENSES

Even though they don't have noses or tongues, moths have excellent senses of smell and taste. They smell through their antennae and taste through their feet.

The fancy scientific words for smelling and tasting are *olfaction* and *gustation*, and every animal in this book has genes that contribute to these functions. Olfaction and gustation genes help the animal sense the presence of certain chemicals, process an understanding of those chemicals, and deliver the information to the animal's central nervous system via olfaction and gustation neurons.

Moths need to smell for predators or airborne particles that might hinder their flight, so their smellers are up top. When they're feeling around on flowers or trees, they're looking for food and safe places to lay their eggs, so they taste their way to survival.

RESPIRATION

Moths breathe through "nostrils," or spiracles, that are located on their "butts," or abdomens. The air tubes, which connect to air sacks, are called trachea, just as they are in vertebrates.

METAMORPHOSIS

Many insects go through metamorphosis; moths and butterflies just do it the most spectacularly. Moths hatch from eggs in their larval form: caterpillars. Then, in the pupal phase, a moth spins itself into a cocoon and a butterfly encases itself in a chrysalis. The pupa breaks down its whole body into basic genetic building blocks, leaving only its respiratory system and a bit of its nervous system intact, and rebuilds itself anew: body system by body system, trachea first and on outward.

The color and patterning on peppered moths helps camouflage them in their surroundings. The same is true for peppered moth caterpillars. Grown-up moths mimic the aged bark of adult trees, and their young caterpillars have bodies the shape and texture of twigs, varying in color from green to brown.

LESSON OF THE PEPPERED MOTH: Here again, the changes that come with evolution have been recorded in real time, and a possible explanation has been isolated in the lab. Darwin thought such changes happened by coincidence, and then by selection, over time. We now see that they may sometimes happen more quickly than this. Part of the reason for this quickness might be transposons.

TERMS DEFINED: Transposons, coevolution

DAPHNIA
(Daphnia pulex)
AND NEMATODE
(Caenorhabditis elegans)

Here we have a *Daphnia* and a particular roundworm (aka nematode) best known as *C. elegans*. You may have poked these see-through buddies in high school biology class, and that's no coincidence: they're model organisms.

As model organisms, both *Daphnia* and *C. elegans* have become key players in the study of transposons. *Daphnia*, for instance, have unusual features compared to a lot of aquatic invertebrates, so it was relatively easy to figure out which proteins in its short genome con-

trolled which of the unusual features. The DNA that didn't appear to have a direct correlation with a trait, that's the stuff that researchers have been messing with lately, to see the differences in non-coding DNA in different generations of *Daphnia*.

ANTENNAE: two pairs, the second of which is used for swimming.

EYE: just one, compound, separate tissues that fuse into one by adulthood.

EGGS: kept in the back, just under the carapace (outer shell).

SHELL SPINE: pokey and useful in self-defense. Multiple spines along the back is one of the inherited traits studied by *Daphnia* scientists.

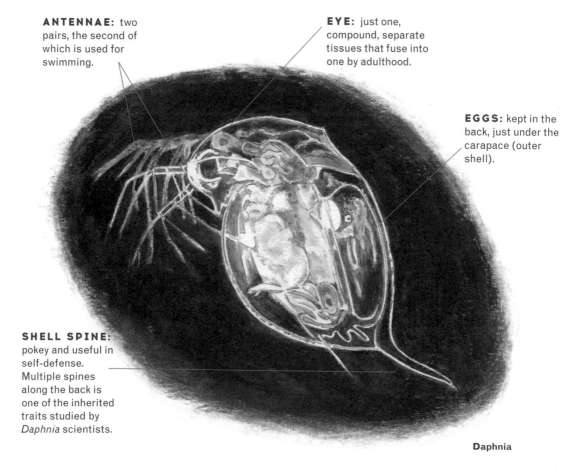

Daphnia

MODEL ORGANISM

Also known as model species, model organisms are animals, plants, bacteria, or any living things that wind up getting studied a lot. Usually these organisms are small (so as to fit in a lab), and they reproduce fast and often, such as yeast or fruit flies. If the model animals are transparent to boot, so much the better. The easier an animal is to propagate and study, the more likely it is to become a model organism. And the more likely it is to be studied a little, the more likely it is to be studied a lot. It's easier to build on existing genetic knowledge than to start from scratch with each new study. Collectively, science knows a lot about model organisms—not because such organisms are more important than others, but just because it kind of worked out that way. (It's a metaphor for evolution, shhhh...)

TERMS DEFINED:

Model organism

SCALE
Daphnia: 0.2–5 mm (0.01-0.20 inches) in length
C.elegans: About 1 mm (.04 inches) in length

FAM·O·METER

60%?

As model organisms, both of these animals are studied for the express purpose of better understanding other animals, especially humans. This process is ongoing. But look at the flip side: While humans have many more genes than *C. elegans*, as much as 00% of *C. elegans*' genome may show up in humans in some form or another.

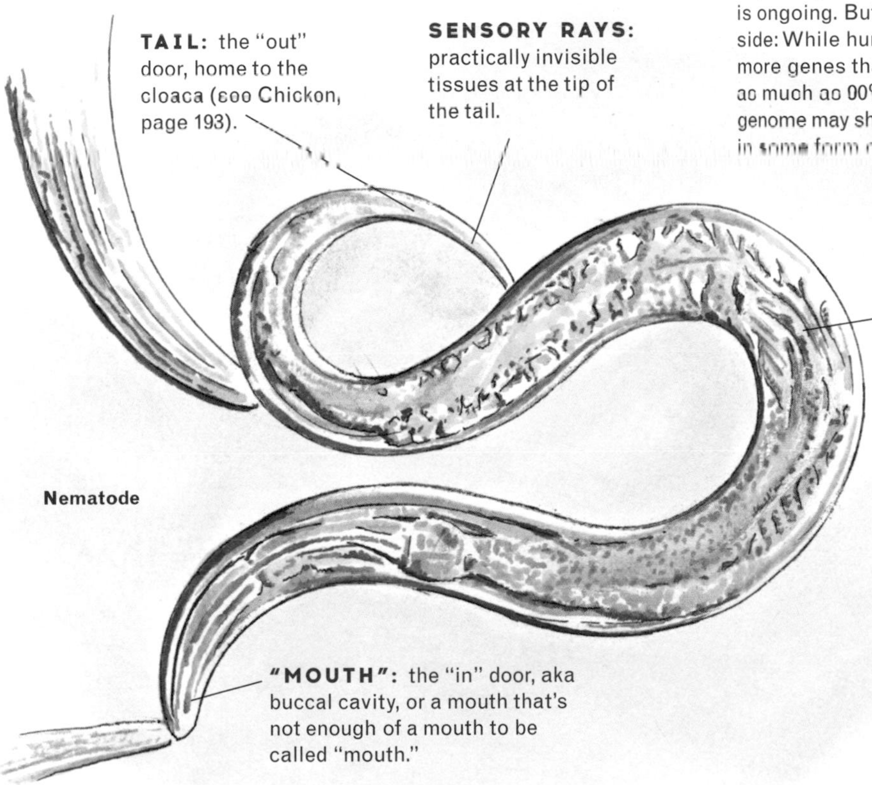

TAIL: the "out" door, home to the cloaca (see Chicken, page 193).

SENSORY RAYS: practically invisible tissues at the tip of the tail.

TESTES: aka reproductive bits, which take up a lot of your body when your body doesn't have much going on.

Nematode

"MOUTH": the "in" door, aka buccal cavity, or a mouth that's not enough of a mouth to be called "mouth."

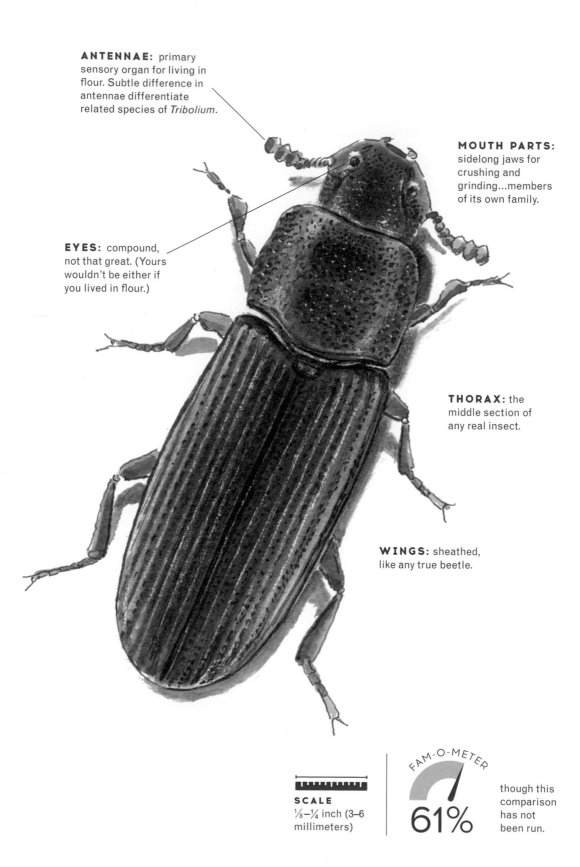

ANTENNAE: primary sensory organ for living in flour. Subtle difference in antennae differentiate related species of *Tribolium*.

MOUTH PARTS: sidelong jaws for crushing and grinding...members of its own family.

EYES: compound, not that great. (Yours wouldn't be either if you lived in flour.)

THORAX: the middle section of any real insect.

WINGS: sheathed, like any true beetle.

SCALE
⅛–¼ inch (3–6 millimeters)

FAM-O-METER
61%

though this comparison has not been run.

RED FLOUR BEETLE

(Tribolium castaneum)

CUTTING OUT THE MIDDLEMAN

The flour beetle is famous in evolutionary biology circles for being a great example of group selection. Yep. Self as predator and prey.

WHEN IN (PRE)HISTORY/ WHERE ON THE RIVER

The first proto-beetles emerged over 300 million years ago and survived the "K-Pg" mass extinction that wiped out 70-80% of life on Earth (see page 169). Today, there are 380,000 known living species of beetle. Living beetles account for approximately 40% of insects, and a full 25% of known species on Earth. This means two things:

(1) being a beetle *works*, and (2) some of those beetles are going to have filled some pretty far niches to survive.

A 2018 Japanese study of the red flour beetle's genome found that this single species can trace its lineage back to Japan, Thailand, and Canada. The red flour beetle is a jet-setter.

360	250	MILLION YEARS AGO		65	0
PALAEOZOIC		MESOZOIC		CENOZOIC	

RED FLOUR BEETLE

...About 200 million years ago, they split from the lineage that would eventually include ladybugs and longhorn beetles.

...About 235 million years ago the flour beetle's lineage split from the lineage that would eventually include scarabs and the mighty Hercules and stag beetles.

...About 123 million years ago, the flour beetle's family emerged, one of the most diverse families of beetles, boasting around 20,000 known species. They're commonly known as darkling beetles because, like the titular freaks in Whodini's hit 1984 song, they only come out at night.

BEASTLY BREAKDOWN

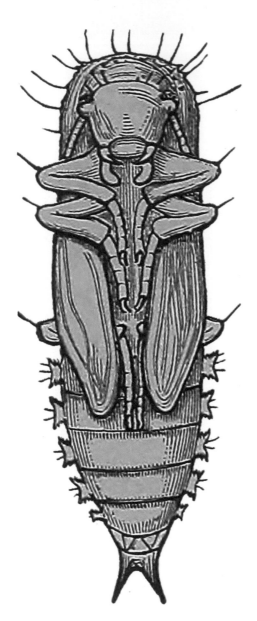

Larva of the red flour beetle, belly side up. From the New York State Museum bulletin, 1916.

SHEATHED WINGS

When most of us think "beetle," we picture an insect with a round carapace (shell) like a ladybug or scarab, but beetles can have longer abdomens and even large heads, as long as they have two sets of wings, according to old-school taxonomy. One leathery set overlays the second as a protective cover. Compared, for instance, to the wings of a "true bug" (like a kissing bug, aphid, or cicada), a beetle's wings meet in a straight line down their back, instead of overlapping in an X.

MOUTH PARTS

A beetle must have two large sidelong jaws called mandibles for crushing and grinding, as compared, for instance, to the proboscis-like beak of a true bug, which is used for poking and sucking.

METAMORPHOSIS

Beetles, too, start as eggs and turn into larvae and then pupae before they become adults. Bugs, on the other hand, emerge from the egg fully formed as it were, looking like miniature versions of their parents, and eventually skip the pupa stage by busting out of their old skin directly.

KIN-CANNIBAL CONUNDRUM

The red flour beetle presents a long-standing debate in evolutionary theory: Many experts consider this "eat-the-young" scenario to be a classic example of *kin selection* (selection by family). But why would an animal group evolve to eat itself alive? Maybe it didn't.

In some laboratory experiments, scientists have been able to lower cannibalism rates in groups that *aren't* very related to one another. So it appears that many of these beetles are in fact ruled by some other kind of selection. Or several kinds—as more modern evolutionary theory suggests. Only time and more information might tell the whole story.

BEHAVIOR

Flour beetles do something most animals do not do: they cannibalize their own population. So frequently, in fact, that the behavior has evolved to be a consistent aspect of flour beetle group dynamics and fitness.

Adult males and females of the species both take part in cannibalizing eggs and pupae—produced by other individuals of course—but so much so that the young of their very own species constitute a critical portion of the beetles' diet. As their name suggests, flour beetles live in flour: human-milled flour and some other grains that you'd use to bake with (which is why you always keep your flour sealed and never scoop it with a dirty scoop, according to what my mother taught me). And as you might imagine, a diet of nothing but human flour, especially sterile, bleached white human flour, might not have a lot of nutritional value,

even to a simple animal—not if that's all it eats.

So where else can a flour beetle get a meal? Well, in a container of flour, there are two options. This weird niche diet doesn't seem to make sense in one regard: eating one's own kin isn't great for the kin, is it? On the other hand, the most direct way to replenish your body of the stuff it needs to keep being a body or reproduce and make more little bodies just might be to ingest another body.

Indeed, beetles in nutrition-poor environments (like low-sugar oats) were pretty much guaranteed to start cannibalizing and would do so earlier in their life cycles. And indeed cannibalizing helped them grow as big and healthy as beetles in nutrition-rich environments (like brewers' yeast). Beetles infected with parasites were also more likely to cannibalize. But then, obviously, if the population as a whole isn't discerning when it comes to eating each other or not, they can eat themselves into oblivion.

But perhaps most interesting of all, larvae that were raised close to eggs they were related to (eggs that could potentially be dinner) sometimes went for the eggs and sometimes didn't. It's as though the trait manifested as an option; it's not great for the family if you eat eggs you're related to, but hey, if you gotta, it's good to have the option.

***

LESSON OF THE FLOUR BEETLE: An animal species isn't governed by just one type of selection: types of selection are interrelated and don't lead to the best interest of the animal. Selection just happens. Positive selection happens after a selective pattern has been set into motion. As far as we know, there is no intentionality or grand plan.

TERMS DEFINED: Group selection

GIANT MARINE ISOPOD

(Bathynomus giganteus)

EASY THERE, BIG GUY

It doesn't much matter to an animal where it ends up or how it gets there, as longs as it survives. The giant marine isopod is a massive, very intense-looking carnivorous bug. But not a bug. It's an oddball for its type, proving that even invertebrates can break the mold, from size to sexual selection.

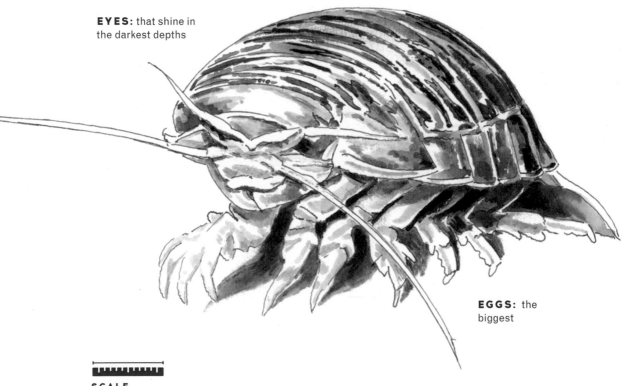

EYES: that shine in the darkest depths

EGGS: the biggest

SCALE
7.5 inches to 2.5 feet
(19 centimeters to
0.76 meters)

WHEN IN (PRE)HISTORY/
WHERE ON THE RIVER

Giant marine isopods are most closely related to the much smaller land-dwelling pill bugs, aka roly-polies aka woodlouse—which are also technically crustaceans, even though they live on land. Which is weirder: that there are tiny crustaceans on land, or that there are lobster-sized ones in the water?

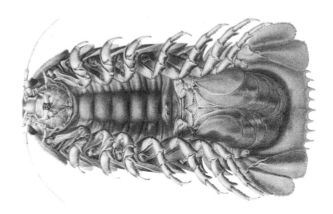

BEASTLY BREAKDOWN

SIZE

It's unusual for an invertebrate to get as big as the giant marine isopod. Even crabs and lobsters don't usually get that big. But this isopod has grown according to a pattern in nature scientists refer to as deep-sea giantism—the same phenomenon that makes the deep-sea giant squid the largest mollusk by a long shot.

EGGS

This isopod's eggs are the largest of any invertebrate—about the size of a quail's egg. Hatchlings emerge from the egg looking like tiny adults, then molt their exoskeletons when they grow.

EYES

This isopod has fixed compound eyes with more than 4,000 individual facets. Compound eyes like this aren't necessarily better for seeing anything; they just sort of happened that way: one eye bud gave way to several, which kept

going. They also have a reflective layer at the back of the eye called the tapetum, which reflects light back through the retina and increases the ability to see at night or under deep murky water. Other animals, such as dogs and cats, have this, too. Humans *don't* have one, which is why our pupils shine red in the flash of a bright light.

SEX

Marine isopods are probably as large as they are because size is a sexually selected trait. For female isopods, size does matter. As such, they have a whole sex-based social structure, with alpha, beta, and "sneaker" males that mate with females behind the alpha male's back.

LESSON OF THE GIANT MARINE ISOPOD: Even invertebrates can be enormous and have weird sex behaviors. That should keep you on your toes.

HUMAN

(Homo sapiens)

100%,
mama. Now seems like
a good time to point
out that we share
around 25% of our
genome with rice,
grapes, and *E. coli*,
and probably all other
living things. (More on
this in Part 2.)

SCALE
From the smallest
man on record,
21.51 inches (54.64
centimeters), to
the tallest, nearly
104 inches (264
centimeters)

BRAINS: too big
for our own good

**SALIVARY
GLAND:** evolved
with a diet of almost
everything

SKIN: some
shade of brown
and universally
sweaty

BIPEDIALITY: pretty
clumsy, actually

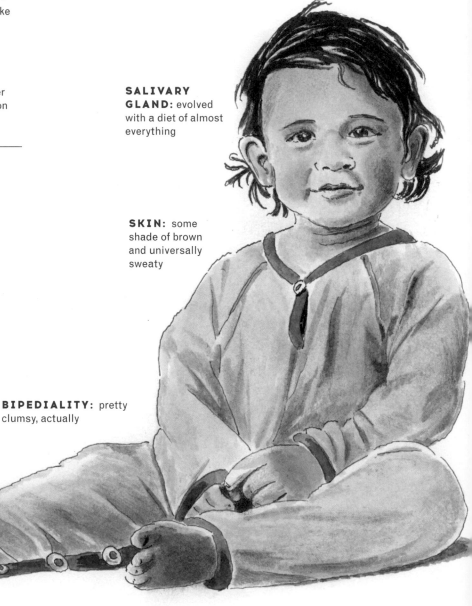

DEFINITELY NOT DESCENDED FROM MONKEYS

Now that we know what Darwin meant by *species* and natural selection, let's find out what he meant by that last part: *"...or the Preservation of Favoured Races in the Struggle for Life."*

To me, that just means "What does success look like?" And something that humans are good at doing is defining things through our own lens.

In 2001, the human genome became the first genome to be fully sequenced. As the first, it became the roadmap for the other genomes to unfurl after it. At first, anyway. But much of our understanding of all genomes—the human genome included—had to wait until enough genomes were sequenced to compare them to each other. And when that happened, we were rather shocked to learn that the human genome wasn't as unique as we'd assumed.

WHEN IN (PRE)HISTORY/ WHERE ON THE RIVER

As far as we know, 6 million years ago or so, there was a divergence in gene pools: the lineage that would become modern-day chimpanzees split from the lineage that would become modern-day humans. (Humans did not descend from chimpanzees, because chimpanzees as they are today evolved alongside us since then.) Technically, science categorizes modern-day humans as "Great Apes" ("Hominids," family *Hominidae*). The fact we can call ourselves "ape" is what separates us from the apes.

Before that, about 11 million years ago, the proto-ape/humans diverged from proto-monkeys.

Before that, about 16 million years ago, the proto-ape/monkey lineage diverged from proto-lemur/bushbaby lineage.

Before that, about 55 million years ago, the proto-ape/monkey/lemur lineage diverged from proto-mouse/rat lineage.

Before that, about 60 million years ago, the proto-ape/monkey/lemur/mouse mammal lineage diverged from the proto-carnivore (dogs, cats, bears) lineage.

Before that, about 85 million years ago, the proto-ape/monkey/lemur/mouse/carnivore mammal lineage (everyone who would eventually have spines) diverged from the invertebrate lineage (everyone who would eventually have gooey bodies, shells, or exoskeletons).

Before that, about 100 million years ago, the proto-vertebrates-*and*-invertebrates diverged from the proto-fungus, bacteria, virus, archae-bacteria lineage. Together, these vertebrates and invertebrates are called eukaryotes. Everyone else is a prokaryote or archaebacterium (an entirely third category of life, discovered in the early 2000s).

Before that, over 1.5 billion years ago, at a time and place that will likely remain a mystery the rest of human history, a living thing emerged that would give rise to all the other living things to follow. It almost certainly wasn't the first living thing to emerge. It was simply the one that survived and thrived long enough to procreate viable offspring that would themselves procreate (or, more simply, replicate, at that point). Science refers to this unknowable organism as LUCA, for last universal common ancestor, aka microbial Eve.

Where did this "life" come from? How did it come out of nothing? It would have started with the same chemicals that DNA is made out of. DNA is, after all, just a type of molecule. From there, the conditions would have had to be just right: certain chemicals (hydrogen and carbon) would all have to have been present under exactly the right environmental factors (temperature, for instance) to catalyze a change in the proto-RNA molecules from nonliving to living.

In an immensely important 2010 experiment, researcher Craig Venter attempted to re-create the conditions of early Earth in his laboratory. And he succeeded and created—essentially—life from nonlife. But even the above working definitions of life are controversial and filled with a lot of educated guesswork, considering how little evidence we yet have of what Earth's earliest life looked like. As NASA's Mary Voytek once told me, "Ask a room full of 100 scientists to define 'life,' and you'll get 120

An **organism** is any living thing. Animal, vegetable, plant, fungus, bacteria, virus, or archaeon.

different answers…Some people will tell you definitively that a puppy is alive and a rock isn't, but there's a lot of gray area in between."

How do we know what we do know? Genomes. Genomes tell a story of DNA combinations that differ but also tell the story of DNA combinations that have remained the same since all the living things on Earth—since the advent of every mountain lion, ant lion, and dandelion.

What makes us human? Well, technically, human lineage specific (HLS) genes, as a group of American researchers put it in a big, cross-institutional study in 2014. The study compared human genomes to the genomes of several great apes to isolate what (at least in the primate line) sets us apart from our closest cousins (genetically speaking). And we really got to pin down who we are in 2001, when we first sequenced the human genome. Since then, we've sequenced over 60 human genomes and counting, including a Korean person, a Han Chinese person, a Yoruban Nigerian person, a European American female leukemia patient, and several white male biologists. (See also Neanderthal, page 77.)

SWEAT

Okay, I'm just gonna tell you right now that all the gross things about humans are the things that make us, us. One such HLS trait is related to energy storage in the skin and probably

BEASTLY BREAKDOWN

related to human sweating. (Animals with hair all over their bodies don't sweat nearly as much or from as many places as we do.)

SALIVARY GLAND

Weirdly, one of the features that's apparently gotten more genetically complex since splitting with mice and primates is our salivary gland. Presumably, this has to do with the changes in diet associated with cooking food, but considering our proto-human lineage, it's awfully tempting to make a joke here about kissing cousins swapping spit.

REPRODUCTION

Another HLS gene codes for the hormone receptor that helps a woman's ovaries notice that it's time to make babies. If this gene is positively selected (not always a good thing, remember), it can lead to overstimulation in the ovaries or premature ovary failure or cause a woman to give birth prematurely

MALE GENITALS

One of the HLS genes identified in the big 2014 study was related to changes in a sensory area in the male genital region. That penis bone you heard about in the blue whale section? Most mammals have it, but whales don't, and neither do humans. So this particular HLS change isn't unique to humans, but the way that the change may have occurred between ape and human is.

POPULATION SHAPE

Humans are the only species whose population has far exceeded our environment's capacity to sustain it. And not in some tree-huggy sense,

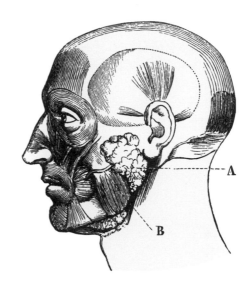

but mathematically. Add up all the resources, and on paper, we as a species shouldn't be alive.

BRAIN

Next to dolphins, humans have the largest brain size for their body mass. But a growing amount of research has shown that the relative size of a brain isn't as important a contributor to "intelligence" as greater connectedness. Ever since we split from mice, humans have seen most of their new genes go disproportionately toward the neocortex, the front area of the brain that we believe controls executive function and higher thinking, like decision-making, reasoning, risk assessment, and social maneuvering.

More massive and complex brains mean more genes devoted to brains, which mean more variation. Old-school thinking might say there are more places for things to go wrong. But as can be seen in *Xenopus* or the avian boom, more genetic variation offers no right or wrong; it simply offers options. And not in a cute "every mistake is an opportunity" inspirational plaque kind of way. In a literal life and death way. Sometimes it goes the way of the mammoth, sometimes the way of the octopus.

MO' BRAIN-Y...MO' PROBLEMS

So far, humans are the only species to exhibit schizophrenia and bipolar disorder, both conditions that appear to originate in the physical brain rather than in hormone-controlling glands and gutty-works; hormone changes can also have a mood-altering effect. Full-on brain problems seem to be a human thing. In a 2018 study, researchers in Europe isolated the gene variations associated with these conditions. Whether or not these conditions are problematic, though, is a societal problem more than a physical one. As with our assessments of intelligent behavior, we've got to be careful not to let our definition of normal make things worse for our fellow humans afflicted by the abnormal.

Unfortunately, what goes up must come down: some of the genes that help us grow big brains have the opposite effect when they get turned off or manipulated. These are HLS genes that are affected by conditions like Zika, where infected fetuses are born with abnormally small brains.

Another area seems to have helped humans develop our uniquely large forebrains, the part of the brain that does all the higher functions, like planning, decision-making, and connecting the dots between various inputs. A related HLS gene codes for a protein in the brain that, on a good day, might contribute higher brain function but on a bad day might cause the serious nerve disorder multiple sclerosis.

Still another HLS gene is a neuron receptor that, when copied too many times, may contribute to attention deficit hyperactivity disorder (ADHD) and even cervical problems in women.

BRAIN AND HEART PLAQUES?

Another HLS trait is a change in a particular protein maker that appears to have shifted from a fully understood "gene" to what researchers call a "pseudo gene." It relates to the buildup and removal of certain biological residues inside the body, and the shift that occurred between apes and humans may contribute to heart disease, Alzheimer's disease, and atherosclerosis.

CANCER: TOO MUCH OF A GOOD THING

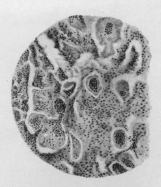

Cancer = cells forgetting to die.

Healthy cells are living cells, but healthy cells also know when to stop. Turn off. Stop multiplying.

Cancer is actually a disorder that happens when cells forget to stop multiplying: when the Off switch is stuck in the On position.

Many animals have specific genes that help assure that this goes well—genes called tumor suppressors. But many generations of humans lack effective tumor suppressors. Another genetic tradeoff, perhaps, but why? And can we trade back?

MEDICAL INTERVENTION

In 2017, when I was at a conference about reproductive technologies, an attendee asked if something like gene therapy was "going against evolution." Whether we're talking in vitro fertilization, DNA manipulation like in CRISPR, or even cancer treatment, it's a slippery slope. What is chemotherapy but a dangerous, untargeted form of gene therapy?

But is it "messing with evolution"? No. This is evolution (which, again, is a word we came up with and are constantly trying to understand). We evolved these brains, bodies, cells, DNA, and everything in between. And we'll continue to do so, for better or worse. In fact, where things look worse, we might skew them toward the better.

COMMUNICATION

Another gene first found in humans has been isolated as playing a key role in human speech. Researchers were able to isolate the gene by comparing genomes among humans with healthy speech patterns and humans with speech problems.

The gene was nicknamed FOX, after an abbreviation of its chemical ingredients. And yes, it brings to mind that terrible Swedish music video "What Does the Fox Say?" whose title is shared by at least three actual scientific articles in actual scientific journals, in case you were wondering.

This communication, experts argue, is the most important job of our big brain, above all else. That it helps us have relationships with one another, which many experts argue is the single most important human-defining trait. Our relationships have given us rituals, memories, and language. Sure, we make and use tools, but even more important is our ability to share the tools among ourselves, in that we make them to help one another and teach one other how to use them. We learn from one another. As is so

THROUGH DARWIN'S EYES

Darwin knew people wouldn't like the idea that humans are related to apes. Even though he never said we descended from them, he knew that's what people would hear. It worried him so much that he left it out of his first edition of The Origin of Species.

Likewise, Darwin didn't love this idea of "fitness" or "fittest" as a sort of value judgment, that success was akin to winning.

A new term has arisen in the genetics age that perhaps he would have liked better. Rather than fitness, some researchers are defining success simply as persistence. Rather than being the fittest animal on the block, you're simply one of many who persists. Any animal—or genome, or gene—that is still around today is persistent. We will all persist as long as our environments allow, and until we cannot.

often the case, the answer here is more easily summed up by quoting science fiction than science: "It's people! It's peoplllllle!" This is unique to humans.

LESSON OF THE HUMAN: What is working for humans today doesn't necessarily mean it's better. There is no fittest; there is only alive, or propagating. There is only **persistence**.

TERMS DEFINED: Persistence

FAM-O-METER

99.9%

But remember: This measure isn't the same thing as the percentage you get back on your DNA test. That measures probable ancestry. The Fam-o-Meter measures *what* your DNA is making with your genome and how much it overlaps with the other animals in this book. If a DNA test were to have two categories, "Same as Neanderthals" and "Not Neanderthal," the result would average between 2.5% and 5% Neanderthal.

SCALE
Females may have averaged a little over 5 feet (155.5 centimeters), males 5.3 feet (165 centimeters).

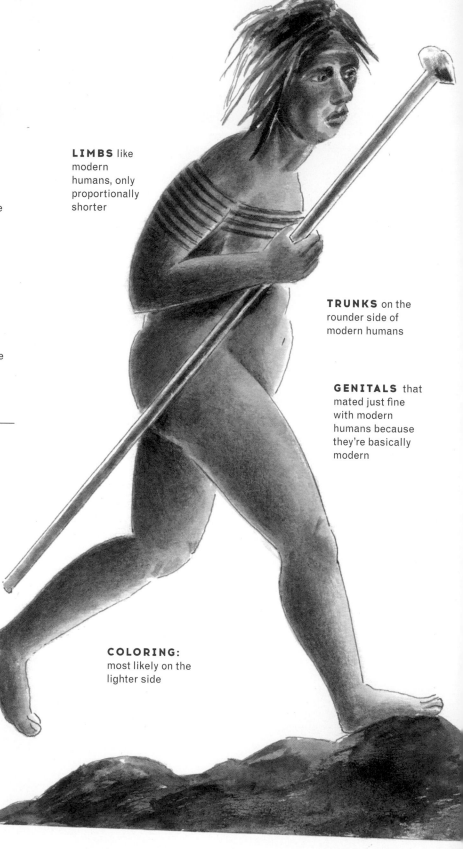

LIMBS like modern humans, only proportionally shorter

TRUNKS on the rounder side of modern humans

GENITALS that mated just fine with modern humans because they're basically modern

COLORING: most likely on the lighter side

NEANDERTHAL

(Homo neanderthalensis)

ACTUALLY PEOPLE

We're pretty sure we understand what makes humans, humans. But *when* did humans become human?

THROUGH DARWIN'S EYES

In 1856, miners working in a limestone cave just outside of Düsseldorf in current-day Germany found pieces of a skull. They could see the fragments were very old and thought at first that they belonged to an ancient cave bear. But this skull was rounder than a bear's, yet thicker than a human's. The valley there was called the Neander Valley, or Neandertal. And so the skull is named.

It was three years before Darwin published *On the Origin of Species.* In his journals, he debated with himself over how to tell the world about his theories. He was thrilled by the discovery of primitive humans and their implications. "It must be admitted that some skulls of very high antiquity, such as the famous one of Neanderthal, are well developed and capacious," he wrote.

But he worried that the idea of "primitive humans" will offend people who believe humans were made in God's image.

WHEN IN (PRE)HISTORY/ WHERE ON THE RIVER

The common ancestor that gave way to both *Homo sapiens* and Neanderthal was alive on Earth—best guess, right now—550,000 years ago. But once they parted ways, did they stay parted?

Neanderthals appeared in the fossil record to have lived 300,000–400,000 years ago. They disappeared possibly the same time modern humans (*Homo sapiens*) appeared on scene. We think.

As the years pass, more humans stumble upon remains from early humans. As Darwin anticipated, modern humans are eager to differentiate between "us" and "them." Even though the scientific community has yet to

agree on the definition of *species* (page 30), they do agree that these squat, heavy-browed humanoids represent a species that is entirely separate from modern man. Fast-forward to the mid-20th century and the popularizing of Ernst Mayr's definition of *species*, and we're still in the clear: it looks like Neanderthals probably spread out over Eurasia 50,000–60,000 years ago, way before modern man, so any species-blurring hanky-panky was surely impossible. Surely.

In Leipzig, Germany, in 2010, researchers sequenced the mitochondrial DNA of the Neanderthal—the Neanderthal from Neander, about 500 kilometers away. This was the big

Fragment of the Neanderthal skull, front view, from biologist Thomas Henry Huxley's book *Man's Place in Nature*, 1863. The book came out five years after Darwin and Wallace published their paper on natural selection, and eight years before *The Descent of Man and Selection in Relation to Sex*, Darwin's book on human evolution.

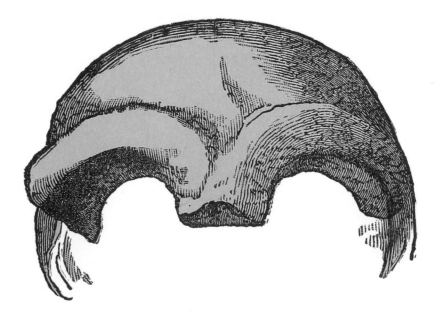

reveal: Would the Neanderthal's DNA look like modern man's? Would it show evidence of interbreeding? Mitochondrial analysis shows no overlap with modern man. By both of Mayr's criteria, modern man is still a discrete species. Modern man continues believing we're the most evolved (see page 71).

Then, in Torquay, England, in 2011, cavers found a jawbone that carbon dated to 40,000 years ago. But it was no Neanderthal: anatomically speaking, it looked like the jaw of a modern human. Hominids shaped like this and Neanderthal hominids all lived at the same time. These remains were all found in Europe, but thousands of miles apart. But if they *had* met, would they have interbred? Back to the DNA-ing board.

The 2010 sequencing was done with young technology—the best available at the time, but not yet good enough to overcome the challenges presented by ancient, degraded DNA, all mixed up with bug carcasses and worm poop and who knows what other biological leftovers from the last hundred thousand years.

Very long story short, over the next several years groups of researchers from all over the world tried the Neanderthal genome several more times. They developed ways to separate out extra insect and worm DNA, they included DNA from a wider variety of Neanderthal remains, and they even figured out a computer simulation to close biochemical gaps in the sequences using proven patterns shared by all hominid DNA.

They compared their findings to modern human DNA from all six human-inhabited continents. Guess what: we interbred. Modern humans have plenty of other DNA that's not identical or shared, but Neanderthal DNA is definitely in the mix. So we did mate. And our offspring was viable, which even Mayr would have to admit makes Neanderthals and modern humans technically the same species.

BEASTLY BREAKDOWN

BODY SHAPE

Neanderthal-specific DNA universally gave them a barrel-shaped chest and proportionally shorter limbs.

RISK FOR TYPE 2 DIABETES

Diabetes risk probably comes from Neanderthal gene variants. They may have adapted metabolic responses to survive periods of famine that simply don't align with modern-day diets.

COLORING

According to a 2017 study, humans alive on Earth today have Neanderthal genes that contribute to hair and skin color, but those colors range from very light to very dark. Neanderthals made it from Africa all the way up into Europe, which means they passed through a range of climates and adapted accordingly: the farther from the equator, the lighter their hair and skin got, just like modern humans. There's no single set of genes that determines hair and skin color, just like in modern humans.

DENISOVAN

Siberia, 2008. Russian archaeologists digging around in Denisova Cave in Siberia found a teeeeny-tiny piece of bone: it literally turned out to be a piece of the tip of a child's finger—well-preserved but so tiny. These archaeologists really knew their stuff. But were not super creative when it came to naming: They called the hominid Denisovan, like the cave.

DNA analysis of the Denisovan remains showed modern humans have Denisovan DNA too, and those who do are most likely to be of Eastern Europe, Asia, and Pacific Islander descent. Modern humans with ancestors from East Africa (modern-day Ethiopia, Eritrea, Djibouti, and Somalia) or southern Africa share another chunk of DNA that aligns with DNA found from a variety of East African fossils (all with different names). And there's a whole other, parallel set of DNA shared by modern humans with ancestors from West and central Africa—but we haven't found any corresponding remains.

So what happened? Early hominids evolved from a less-human-ish ancestor in the middle of what is today known as Africa. About 300,000 years ago, some of them went on the move. One group stayed in Africa (the one with no fossils from the time yet). Some made it up to the horn of East Africa or the cape of southern Africa, some moved into Asia (Denisovans), and some into Europe (Neanderthals).

But wait. There's more.

In 2017, 220,000-year-old Neanderthal remains were found in a different cave in Germany (instead of "cavemen" we should really be calling them *Höhlenmenschen*) that had modern human mitochondrial DNA (mDNA), the same lineage that gave rise to modern humans who stayed in Africa. That means that before all those groups trekked out into Europe, a proto-Neanderthal male mated with a proto–*Homo sapiens* female. She stayed in Africa and made some more babies with *Homo sapiens* mDNA. Their offspring took her mDNA up through Europe along with all the other proto–*Homo sapiens* and proto-Neanderthals and proto-Denisovans and hybrids of proto-all-of-the above. By around 30,000 years ago, the non–*Homo sapiens* DNA was pretty much divided into those four geographic groups: folks with Neanderthal DNA, folks with Denisovan DNA, folks with the early outer-African DNA, and folks with the early-African DNA that was still non–*Homo sapiens*, even though *Homo sapiens* were there, too (and maybe mostly there, at that point).

If you're confused as to where one group begins and the other ends at this point, well, exactly. This all goes back to individuals, the only difference among the groups, between "us" and "them," being who you are most recently related to. Eventually, though, all Neanderthal mDNA was overcome by *Homo sapiens* mDNA. But that 2.5–5% of DNA persists today.

SLEEP PATTERNS

As was proven with the help of the fruit fly *Drosophila* (page 137) and the mouse (page 183), circadian rhythms are something most animals on Earth inherited from a common ancestor born millions of years before Neanderthals and *Homo sapiens*. But the difference between being a morning person and a night person has more to do with when your ancestors needed to be awake, which had to do with sun exposure. Neanderthals, it seems, were more often night people.

TOBACCO ADDICTION

Tobacco probably didn't exist in the Neander-thals' Europe; as far as we know, it first arrived around 400 years ago when early explorers brought it back from the Americas. But chemical addiction in modern humans has a genetic component, some of which could be Neanderthal. Ironically, one of the reasons that Neanderthals died out might be the very reason modern humans could handle the habit of smoking in the first place.

In 2016, American researchers found a genetic mutation linked to a higher tolerance for breathing smoke. Folks who have it are less likely to develop chest infections or lung cancer after breathing smoke, which is full of a ton of smoke-specific toxins and carcinogens. This mutation hasn't been found in ancient Neanderthals, Denisovans, or any other primates. We know that all early hominins made fires, but it seems that some of them could take the heat, while Neanderthals and Denisovans had to get out of the kitchen. And we know that cooking with fire and eating charred meat were early human adaptations that probably contributed to the evolution of bigger, more socially and technically complex brains.

Eventually, enough modern humans survived smoke to try inhaling it on purpose and realized they liked it. Unfortunately, some of that old DNA persisted. If these compound theories are true, here's a perfect example of how evolution doesn't improve; it has no end goal, and it doesn't always make sense. Modern man, more than any other animal, gets addicted to stuff that kills us.

HEAD

You don't need DNA to see that the Neander-thal skull has a low, prominent brow, and Darwin was right: the Neanderthal's braincase was big, though not as big as ours. But don't judge: modern neuroscience tells us that brain size (and even brain size relative to body mass) has very little to do with intelligence. Brains can change in an animal's lifetime, and besides, *intelligence* is a word modern humans made up to measure how smart things were compared to the humans who made the measurements.

MUTATION: ACTUALLY A LOT LIKE IN THE MOVIES

Mutation is simply a fancy word for a genetic change. It's another word that came about before we knew the mechanisms of how such things occurred, so it started out as a reference to a small physical change. Then it referred to a visible change in a chromosome. Now, it means a change in a specific section of DNA, and it can come about in two ways (that we know of).

Hereditary mutations are mutations that we get from our parents: either they have them and pass them to us, or they occur during the process of their parental DNA coming together to form us.

Acquired mutations occur during the lifetime of an organism, and can occur for any number of reasons, many of which science is still struggling to understand: exposure to radiation, extreme illness, starvation, or stress, for instance.

These are moments when DNA is vulnerable, because it's undergoing change. DNA has systems in place to keep the threads from being lost and to tie up loose ends. But accidents happen. A mutation might entail repeating information in a place where it should stop, or stopping in a place before it's due. RNA is a code, so when it's decoded on the other end, a typo gets translated into a real, physical mistake. Like a recipe for cake that leaves out flour, or plans for a building that specify a joist instead of a stud, or a recipe for a fly that indicates "eye" where eyes ought not to be.

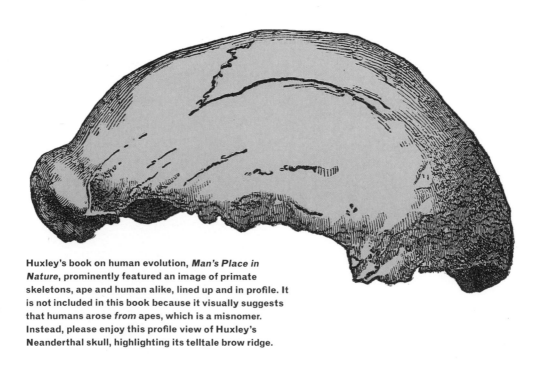

Huxley's book on human evolution, *Man's Place in Nature*, prominently featured an image of primate skeletons, ape and human alike, lined up and in profile. It is not included in this book because it visually suggests that humans arose *from* apes, which is a misnomer. Instead, please enjoy this profile view of Huxley's Neanderthal skull, highlighting its telltale brow ridge.

■ Together, Neanderthal and Denisovan DNA probably makes up 5–7% of human DNA on Earth today.

■ Living humans who have Neanderthal DNA usually don't have more than 5%.

■ An ancient human unearthed in 2004 and dating back to about 1100 BCE showed about 10% Neanderthal DNA.

THINGS MODERN HUMANS LIKELY INHERITED FROM DENISOVANS:

"Functionally relevant" contributions of Denisovan DNA (aka contributions that we can see) include living at high altitude and surviving on little oxygen. Previous studies of Tibetans, who live at elevations of over 9,800 feet (3,000 meters) already showed that they inherit some sort of gene variant that helps their red blood cells, use oxygen 80% more efficiently. And sure enough, they got that from Denisovans.

THINGS MODERN HUMANS SHARE WITH BOTH NEANDERTHAL AND DENISOVAN GENOMES:

Immune cell receptors that look like they're custom-made to protect against very specific viruses/bacteria. "Toll-like" receptors, they're called, like an attendant at a toll booth, ready to stop a specific car for a specific fee.

Allergies. Not any in particular, just a higher likelihood of having them. Thanks, guys.

Genome differences we haven't figured out yet are a more numerical matter. So much of what the Neanderthal IS is seen in the ways it is NOT like we humans of today. And there are some obvious differences between our two genomes. Like 125 places where sequences were totally taken out or totally added, 45 places where genes were spliced together, and 87 places where proteins look different than they should (although six of those might still function the same way without looking the same).

And three of *those* differences have to do with the way chromosomes divide and redefine, which is a part of the body's cell processes that scientists *used* to think shouldn't change over time. But lately, they're finding out they're wrong. And as you might imagine, rapid changes to the way a cell divides just might mean rapid jumps in evolution. Just ask the octopus (page 255).

LESSON OF THE NEANDERTHAL: Humans are hybrids, too.

TERMS DEFINED: Mutation

BOTTLENOSE DOLPHIN

(Tursiops truncatus)

SCALE
6.6–13 feet long
(2–4 meters)

FAM-O-METER

70%

of your genes overlap with the dolphin's. Considering how long ago we diverged on the river of life and how sophisticated the dolphin has become since then, one might expect this number to be lower. But the dolphin is still a mammal. Plus, some of the directions in which the dolphin has evolved overlap with the ways we humans have evolved.

BRAIN: big and complicated

BLOWHOLE: actually a nose

FACE: an echolocating machine tipped with a deadly weapon (rostrum)

TAIL: powerful and efficient—elongated, fluked, and basically one big muscle

ALL IN THEIR HEADS

Like humans to bonobos, most of a dolphin's genes are shared with other cetaceans (whales and dolphins). But—again like humans and bonobos—the genes that make a dolphin special appear right where it counts: in the brain. The same groups of genes appear to have separated the elephant from the manatee, the crow from the chicken, and the octopus from the clam. Also, dolphins are apex predators, so take all that plus mad hunting skills and you've got an oceanic wolf with the social skills of a human movie star.

WHEN IN (PRE)HISTORY/ WHERE ON THE RIVER

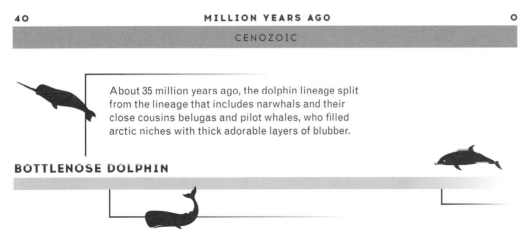

40	MILLION YEARS AGO	0
	CENOZOIC	

About 35 million years ago, the dolphin lineage split from the lineage that includes narwhals and their close cousins belugas and pilot whales, who filled arctic niches with thick adorable layers of blubber.

BOTTLENOSE DOLPHIN

Picking up after the split between baleen whales and toothed whales, sperm whales split off from the rest of the toothed whales about 32 million years ago. They've since evolved to be the only really large toothed whale, filling what must actually be a pretty darn narrow niche: hunting squid, the gianter the better.

Even though the advancement of genetic science is mixing up the classification game, there are roughly 60 other species of dolphin, of which the bottlenose is just the most famous, thanks to being immortalized in TV, film, and dolphin parks everywhere. Dolphins range from the diminutive, tricolored, and poorly named common dolphin to the orca (aka killer whale), which can get up to 18 feet (5.6 meters) in length. Bottlenose dolphins went their own way about 5 million years ago and have become the most populous dolphin species, living throughout most of the world's oceans.

BEASTLY BREAKDOWN

SONAR (AKA ECHOLOCATION)

The sonar system of a bottlenose dolphin is, as far as we know, the most sophisticated animal sonar on the planet. Like bats, dolphins emit sound from vocal cords, then use the returning sound to paint a picture of what they're seeing. And like bats, special genetic markers for echolocation show up in their genome. A bat isn't a bat and a dolphin isn't a dolphin without sonar.

This trait likely didn't come from a common ancestor of bats and dolphins, since most intermediary animals don't have sonar. Those that do have it show different variations in the genome, depending on the sophistication level of their echolocating ability.

A dolphin's sonar is so sophisticated, it's as though they have a sixth sense (or rather a second fifth one, as they don't really smell underwater so much as they taste). This sense is so powerful, it can even detect shapes that are beneath the surface of whatever it is they're aiming at, like fish hiding under sand or vital organs inside the body of their mortal enemy, the great white shark.

ROSTRUM (SNOUT)

Dolphins' refined rostrums help them zero in on whatever it is their echolocation has allowed them to pinpoint, be it fish or enemy shark bellies. The dolphin's hard, strong rostrum serves as a weapon, which they use to ram into sharks with enough force that they can kill a great white twice their size.

BODY SHAPE

Unlike most baleen whales, snub-nosed porpoises, slow-moving arctic pilot whales, narwhals, and belugas, dolphins are built for speed. Their elongated rostra and smooth skin keep them hydrodynamic for top-speed fish hunting and maximum fun.

BRAIN

The brain and nervous system are perhaps what most make a dolphin a dolphin. A 2012 study compared dolphin genomes with others in the cetacean family and found big changes in 228 of the genes, all related to brain size, number of neurons, and most notably interconnectivity. This last piece—the brain's white matter, which helps the brain make connections across itself, versus gray matter, which seems to store information and control the rest of the body— seems to be the culprit for dolphins' particular intelligence, which can best be described as "social intelligence." For example:

■ Researchers have found that dolphins seem to communicate with one another vocally before launching a coordinated attack on a group of prey.

Play: Another way that humans identify intelligence is in recognizing when animals are engaging in activities that don't serve a particular purpose related to survival. This, scientifically speaking, counts as play.

If it wasn't obvious enough that dolphins play during a lot of their waking hours, photographic footage of a dolphin playing with a humpback whale is awfully hard to argue with. The young dolphin would swim up to position itself atop the humpback's rostrum, then slide down it, again and again.

THROUGH DARWIN'S EYES

In their travels through the tropics, Darwin and Wallace would find it hard to avoid dolphins, which are abundant the world over and love to dive in and out of the wake of ships. But although dolphins and porpoises were well known, they were not well studied, and the two families were more or less interchangeable to observers.

Darwin was still marinating his "Big" idea after returning from his trip, but he did begin to publish some of his direct findings in a series of publications that would eventually become *The Zoology of the Voyage of H.M.S Beagle...*[title significantly truncated] *During the Years 1832 to 1836*.

Darwin helped edit the book to include tables of his that carefully cataloged observations and measurements. Each entry ended—not unlike the entries in this book—with a paragraph of his commentary. The commentary for the single dolphin (porpoise) entry reads:

"This porpoise, which was a female, was harpooned from the Beagle in the Bay of St. Joseph, out of several, in a large troop, which were sporting round the ship. I am indebted to Captain FitzRoy for having made an excellent coloured drawing of it, when fresh killed, from which the accompanying lithograph has been taken."

The drawing snubs the dolphin's nose a bit (which actually does make it look like a porpoise). But it captures the unique coloring that identifies it as what we now know as the dusky dolphin or FitzRoy's dolphin.

Darwin was the first to declare the new species, which he dubbed *Delphinus Fitz-Roy*, in honor of his captain's captaining, drawing skills, and overall contributions to natural history. The dolphin's official scientific name was later changed to *Lagenorhynchus obscurus*, but the origin of the discovery of the species of Fitz-Roy's dolphin was a team effort.

- Dolphins seem to have vocal patterns that correlate with individual members of a pod. I hesitate to call it a name, but…you know, kind of like that.

- Dolphins, like elephants and apes, have an acute reaction to death and will spend days if not weeks carrying around the bodies of dead pod members, especially calves.

- Dolphins remember other individual dolphins they've only met once before after years apart.

HANDLESS TOOLS

Dolphins don't have hands, but they have been recorded making and using tools. One group of bottlenose dolphins would use pieces of sea sponge to blot along the sand on the ocean floor, feeling for different types of sea creatures while both protecting their rostrums and disguising their noses from wary would-be prey. Even more interesting was that of the two local populations of dolphins, one group was using the tool, and the other wasn't. This sort of socially based information counts, by definition, as **culture**.

LESSON OF THE DOLPHIN: By human standards, dolphins could be our aquatic intellectual equals. But a success as arbitrary as intelligence looks different when everything about the animal's lifestyle is utterly different than ours.

TERMS DEFINED: Play, culture

NECK: "missing" a few vertebrae, having adapted them for use in its back

EYES: not so great

WHISKERS: highly sensitive, helpful for finding food at the bottom of a muddy bay

BODY SHAPE: like dolphins and whales, just less hydrodynamic

MANATEE

(Trichechus manatus)

CURVES IN ALL THE RIGHT PLACES

The manatee (*Trichechus manatus*) is a marine mammal of very different lineage than whales, dolphins, and porpoises. Yet its body has a cetacean-like shape. This is no coincidence, nor is it a matter of family inheritance. A 2015 study compared the genomes of a manatee, an orca (which is a type of dolphin), and a walrus. The study's authors expected the three different body types to have three different genetic recipes. To their surprise, the body-shape-making genes in all three animals were practically the same.

This is another example of convergent evolution, but with an interesting implication: convergent evolution is so defined because the animals with a similar trait haven't inherited that trait through the same means. But if this finding about the manatee holds true, it means that convergent evolution can itself cause genetic similarities in an animal. Those overlapping genes still won't have come from a common ancestor, but they'll look the same in DNA form all the same. Another twist in the river of life.

LESSON OF THE MANATEE:
Inheritably disparate but characteristically convergent traits can cause convergent patterns in animals' DNA, too.

AMERICAN CROW

(Corvus brachyrhynchos)

"FLYING MONKEYS"

As in the human and the dolphin, the defining feature of the evolved crow is its intellect. Some researchers call crows the chimps of the sky, what with their big brains and ability to make and use tools. They provide another view into our understanding of intelligence. But are they so smart that they're stealing what's ours?

BRAIN: big and complicated

EYES: superpowered and hyperconnected to that big, complicated brain

BEAK: might as well be hands

SCALE
About 17–21 inches (43–53 centimeters) long

FAM-O-METER
73%

This is a conjecture based on the known overlap between humans and chickens. Humans and chickens share a lot, but humans and crows may share more.

WHEN IN (PRE)HISTORY/ WHERE ON THE RIVER

Crows are part of a newer group of birds called corvids, which emerged from the rest of the bird lineage about 11 million years ago. It includes crows, ravens, rooks, and magpies.

It may be important to remember that crows and humans *did* share a common ancestor, but that was more than 50 million years ago.

BEASTLY BREAKDOWN

BRAIN

Relative to body size, a crow's brain is comparable to a chimp's. But remember: size isn't everything. By another measure, the number of neurons in a crow's forebrain is relatively equivalent to a chimp or bonobo. And perhaps most important, their brains have that same cluster of genes that seems to relate to neural connectivity (and therefore our human definition of intelligence) as is found in other social animals like humans, dolphins, and elephants.

VISUAL RECALL AND COUNTING

The crow's brain seems to do best when faced with tasks that connect the eyes to the brain and memory. Indeed, its rather large genome shows a lot of additions around genes related to visual processing. Its party tricks include:

- Recalling up to 100 places where it has hidden food

- Recalling which of those hiding places has the most perishable food and returning to those locations first

- Recognizing facial details in humans and apparently disseminating information about the trustworthiness of clocked humans

- Dropping hard-shelled nuts into a crosswalk so that passing cars can crack them open and the crows can safely retrieve the nuts when the Walk sign illuminates

- Mental template matching. In a 2017 study, a group of Australian researchers set up an experiment wherein one group of crows got a reward from a faux vending machine. They quickly learned that only one size of a token for the machine would earn them a reward. Somehow, they would share information with other crows that did not have experience with the vending machine. The other crows would show up, manufacture their own token according to some shared instruction, and get a reward from the vending machine.

PLAY

Another indication of a rich internal life is whether an animal seeks out play. Crows have been recorded sliding down the snowy slope of a roof, over and over again. It does look, scientifically speaking, really fun.

PARENTING

It turns out that parenting is as much a bird trait as it is a mammal trait: attentive parenting goes back as far as the advent of the archosaur lineage back in dinosaur times. A 2017 study found that even the manner in which a bird flies—the actual movement of their wings—comes from parental teaching. You might think the way a bird uses its wings correlates more with the shape of its wings or what and how it eats and nests, or even instinct. But it turns out baby birds really do need to be taught how to fly.

TOOL USE

Not 30 years ago, the average person believed that humans were the only animals on Earth

The Northwest crow (*Corvis caurinus*), a species slightly smaller than the American crow, evolved to fill a specific niche: the shoreline of North America's Pacific coast. Illustrated dining on a crab claw; Louis Agassiz Fuertes for *Bird Lore* magazine, 1919.

that could use tools. Now we know that apes, monkeys, dolphins, octopus, and even some clever house cats use tools to make their lives easier. But crows really take the animal tool trade to another level:

- A type of New Caledonian crow not only used sticks to hook bits of food, it would modify wire, bending it into various shaped hooks to be best suited to the job.

- A species of crow in Hawaii would take the hard, spikey leaves of a certain tree and strip off various parts of the leaf to make a variety of tools to dig into hard-to-reach crevices and hook insects and grubs.

SOCIAL LEARNING

All of the above, from tool making and trick sharing to identifying which humans are not to be trusted, show evidence that crows can communicate among members of their communities and even pass knowledge down through generations. The mechanisms by which they do this are not yet clear. But compare these achievements to traits we've long considered uniquely human, and the list of HLS intelligences seems to dwindle all the time.

How do we know all this? Most of the findings here were the result of behavioral studies: observation and experimentation. But a recent study examined the crow's genome using several sequencing techniques at once. Usually these studies rely on one technique, and the researchers (ideally) are careful to draw conclusions only from the information the single technique allows. (This doesn't always happen: Scientists get excited, too, and sometimes they overstate their findings. That's why scientific papers have sections called "Limitations" and "Recommendations for Further Study" and why scientific journals are peer reviewed.)

The researchers behind this 2017 study, though, decided that there was wisdom in finding the right tool for the right job. They sequenced a Eurasian crow's genome with several different independent technologies (one of which is called SMRT). This study's findings definitively showed that crows had large genomes (and therefore a lot of genetic variation to choose from). At the same time, it showed that closely related crow species would keep to themselves and not interbreed, even if they had overlapping ranges. Crows of a feather flock together.

LESSON OF THE CROW:
Intelligence? There's more than one way to crack that nut.

AMERICAN PLAINS BISON
(Bison bison)

SON OF A BEEFALO

The story of the cow and the bison is a parable in success. Thanks to humans, the cow has become one of the most successful animals on Earth. The bison, however, was nearly hunted to extinction by humans. But then brought back from the brink by humans, thanks to cows. Why are humans all up in bovine business? Food.

HUMP: from enlarged vertebrae, like the giraffe's neck

HORNS: not the same as antlers

SCALE
1,000–2,000 pounds
(453.5-998 kilograms)
at the shoulder

COW

(Bos taurus)

SCALE
324–3,005 pounds
(147–1,363 kilograms);
height 4–4.8 feet
(122–137 centimeters)
at the shoulder

FAM-O-METER

80%

Cow genomes have a large overlap with humans—so much so that early studies seemed to suggest that humans and cows overlap even more than humans and mice, even though humans and mice are more recently related. This turned out not to be the case, but the overlaps we do share are likely related to the fact that we raise them to sustain our own bodies.

SHARED TRAITS BETWEEN BISON AND COW

Size: Bovines are on the larger end of the prey animal spectrum.

Herding: Herding behavior in both wild bison and cattle is a now instinctual adaptation that helps keep them safe from predators.

Horns: Horns are not just less complicated antlers. Antlers are made of bone, whereas horns are made of hard, fibrous keratin, the stuff hair is made of.

MEAT: made of the same proteins as human meat and bison meat

MILK:
OG GMO

AMERICAN PLAINS BISON

(Bison bison)

WHEN IN (PRE)HISTORY/ WHERE ON THE RIVER

Cows and bison are both bovine, members of a specific branch of the ungulate rivulet shared with water buffalo.

All of the bison alive today descended from the 600 left in 1894. Just 1,000 years before, North America had been so rich with bison that they roamed in the thousands.

BEASTLY BREAKDOWN

HEAD

Bison heads are bigger and broader than their cattle counterparts. One useful application of the head is that they use it to push around snow in the wintry months on the plain, making way for their footfall and rooting up dormant plants to keep them fed without having to migrate to warmer climes.

"HUMP"

In general, the bison has a hump on its back that looks like a camel's fat and serves a similar purpose. However, it is actually the product of elongated vertebrae, much like the giraffe, only elongated in a different direction.

COW BITS

When the bison population tanked in the mid-1900s, crafty mid-century ranchers realized they had to start breeding the remaining bison quick, or lose them altogether. Most of the American bison left were free-ranging, but a few hundred that ranchers had gathered up in Yellowstone National Park were corralled, single, and ready to mingle. With cattle DNA in the mix, the bison population soon recovered. But then again, it wasn't strictly a bison population anymore. Today, cow DNA is being slowly bred out, but most every bison you'll see, wild or otherwise, is still the progeny of that Yellowstone/cattle matchmaking.

LESSON OF THE BISON: It's better to be a hybrid than to be extinct.

COW

(Bos taurus)

THERE FOR US

WHEN IN (PRE)HISTORY/ WHERE ON THE RIVER

What makes a cow a cow? It seems that around the time the first cattle lineages split from their water buffalo relatives, they started having a lot of repeated sections of DNA. These repeated elements meant that their genome became larger over time, which means that there was more genetic material there to work with than in an animal with a shorter genome.

BEASTLY BREAKDOWN

HEIGHT

If you've done any international travel, you'll have noticed that cows' body shapes differ depending on where you are on the globe. In 2018, a massive Australian study compared genomes of 58,000 head of cattle from around the world. The first trait they looked at was height. Height, as we saw in the giraffe, is complicated: it depends on a great number of genes and genetic chunks, not least of which are the length of leg bones and neck bones. Once the researchers organized the cattle by height, they used what they learned as a guide to dig into the other traits.

METABOLISM

Most of the cow's metabolism-related genes look pretty much the same as metabolism genes in other mammals. There are a few notable differences, though: five genes that have been deleted completely or look very different from similar genes in humans. We're learning more about these genes all the time, but the short version is that cows and humans digest things very differently, on a molecular level. This makes sense, because we don't eat grass or have four stomachs.

MILK

The cow isn't the only mammal that makes milk. By definition, literally every other mammal makes milk. But why is it the most famous milk maker? The cow's genome has more variation than other mammals in genes related to milk making. It shares this trait in common with other animals domesticated—at least in part—for their milk: goats, sheep, and even horses. But the milk part of the cow genome is more complex than any other that we've seen yet. Is this because humans have been breeding cows for milk production for millennia? Probably. Cows that make more and better milk get to pass their genes along, making the milk genes of their offspring that much richer. Humans have been genetically modifying other organisms since before we even knew what genes or genetics were.

MUSCLE

Similar to its milk situation, the cow genome also has more variation than other mammals in genes related to muscle proteins.

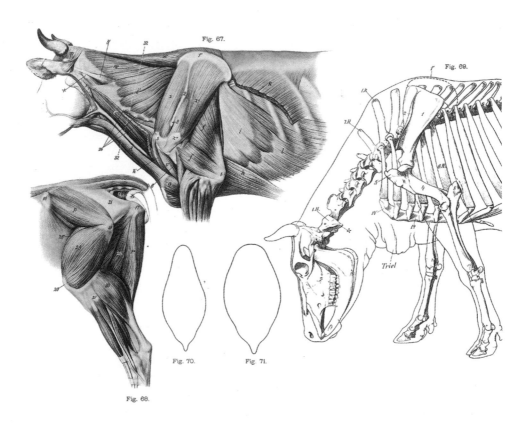

Partial illustrations of cow musculature (beef), udders (without milk and with), and skeleton. By veterinary illustrator Hermann Dittrich, 1889.

PROTEIN

We've learned that DNA codes for proteins and enzymes (via "genes"), and also contains other directions for "how to make a living thing" (via transposons). But what *is* a protein?

A protein is a large molecule—a bunch of atoms stuck together, the simplest form of a chemical. It is a biomolecule or biochemical, meaning it is the type of molecule that only shows up in living things. And a protein has a special job: to make things or make things happen in a living body. Proteins code for many things, from transmitting information along nerves to kicking off digestive processes, but perhaps the easiest proteins to understand are muscle proteins. These are the proteins we're familiar with ingesting when we're hitting the gym to get huge. These proteins tell the body: "Make more muscle."

Our muscle proteins are similar enough to cow proteins that when we eat cow meat, *their* proteins actually tell our bodies to make more muscle. And our proteins are similar because they came from a common ancestor. Evolution is the reason we can build muscle from eating meat.

IMMUNE RESPONSE

The cow genome also shows more variation in genes related to immune response, which also likely has to do with human propagation. Farmers choose to breed the animals that seem the heartiest and whose health allows them to grow big and produce milk and calves. It could also be a trait that led to humans domesticating them in the first place, all over the world, on every continent where humans live.

LESSON OF THE COW: Sometimes you have to get eaten to get ahead.

TERMS DEFINED: Protein

NEAVES WHIPTAIL LIZARD

(Aspidoscelis neavesi)

VIRGIN BIRTHS

The whiptail lizard isn't the most unusual-looking reptile you'll ever see. In terms of epic evolutionary story fossils, like dinos and their mass extinction, it's probably the most boring one in this book. But recently, it's gotten a whole lot more interesting. It has started reproducing its genome in a way that's so successful it's practically cheating.

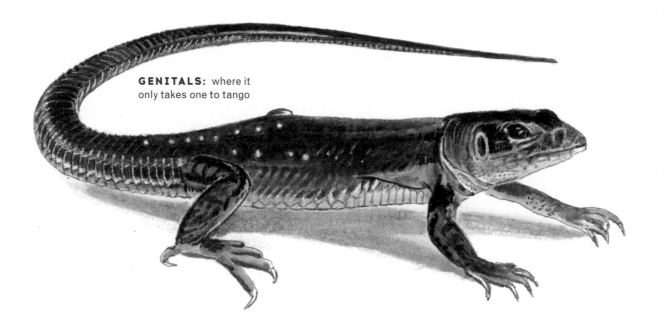

GENITALS: where it only takes one to tango

SCALE
2.3–3.1 inches (60–80 millimeters), not including tail. In finding this fact, I learned that sometimes lizards are measured by SVL, snout–vent length. Dibs on Snout Vent as a band name. #sciencebandnames

FAM-O-METER
67%?

The genetic crossover between humans and reptiles and humans and birds is hard to estimate unless the two genomes are compared directly. As we'll see, there are just too many factors at play.

WHEN IN (PRE)HISTORY/
WHERE ON THE RIVER

Despite their erroneous association with dinosaurs, reptiles seem to be famous as having been one of the oldest animal groups on Earth. In fact, the first reptiles began to emerge around 320 million years ago, pre-dating no one but the mammals.

According to a 2005 comparative genomic study, whiptail lizards are part of a large reptile group characterized by forked tongues and scaly skin. This group includes snakes, skinks, and Komodo dragons.

The Neaves whiptail, specifically, has been around only since the mid-twentieth century. That's when American researchers created the new species in their lab by breeding two existing species of whiptail lizard (at that point known only by their Latin names, *Aspidoscelis exsanguis* and *Aspidoscelis inornata*). Usually, breeding just a few lizards doesn't become a whole new species. It would just become a new generation, and a hybrid generation at that. If the hybrid generation were lucky enough to be viable (as opposed to being sterile, which many hybrids are), they would just go ahead and continue breeding with other existing species and maybe eventually shift into a new "species," after many, many more generations. Because a new generation lizard couldn't only interbreed with itself too…Right? It would become too interbred to be viable the next generation around. Unless…

Female whiptail lizards do reproduce asexually, meaning without males, meaning they clone themselves. But a whole population doesn't usually survive as clones alone—they need a little genetic diversity to keep the population healthy. And somehow, this newly evolved "species" of whiptail lizard has managed to reproduce offspring who are clones… But clones with different genomes than their clone-mothers. It's… Complicated. Terms like "species" and "clone" barely apply.

REPTILE: YOU KEEP ON USING THAT WORD…

Before we had genomic testing, or even genetic testing, or even very many good fossils to compare, *reptile* used to be defined as cold-blooded vertebrates with dry skin (as compared to wet) that lay their eggs on land (even if they spend a lot of time in the water, like turtles and crocodilians).

But ever since we dug up the connection between birds, crocodiles, turtles, and dinosaurs, collectively called *archosaurs* (see the next section, on hoatzin), the idea of "reptiles" has started to feel as outdated as "species." While turtles and crocodilians are still considered "reptiles," the newer groups of reptiles (a little like the newer birds and newer fish) might end up being the only reptiles our grandkids learn about.

SEX DETERMINATION

Scientists have long known that animals reproduce in a number of ways, not all of the ways straightforward. Usually, it involves a male and a female, but some animals (like banana slugs) can reproduce hermaphroditically, meaning they have both male and female sex organs and can do the job themselves. Some animals, though not many, reproduce asexually, also called parthenogenesis. This means that a female animal can reproduce without the help of a male. How is this possible? Well, the first question is, "What do we mean by *sex*?" Or rather, "What do we mean by *male* and *female*?"

In many animals, like humans, sex means sex organs: if you have male parts, you're male; if you have female parts, you're female. But in some animals, the parts don't really look all that different from one another, so the only way to differentiate is to look at the chromosomes. Animals (like most humans) with XY chromosomes are male, and mate with females (XX). Some animals with XX chromosomes (like grasshoppers) are female, and can also produce males with just one X chromosome. Animals with ZW chromosomes (like chickens) are female, and mate with males (ZZ).

But in some animals, females have multiple chromosomes in a variety of arrangements. Female cloning wasps, for instance, have 32 chromosomes, while males have 16. This is because the parent started out with 32 chromosomes and either cloned them to make more females, or used half of them to make males.

In every case, *female* is so named because it is working with *more* DNA, more or longer chromosomes. And that is, by definition, why only females can reproduce asexually. Ideally, an animal starts out with more than one set of chromosomes because there's more to rearrange, which is a huge part of how evolution works. The parents' DNA gets transformed, becomes a new arrangement of DNA, their offspring's brand-new, never-been-seen-before reshuffled combination of DNA that will grow into their baby. Eventually, that baby's DNA will go out and meet some other animal's DNA, and this gets really messy really fast unless there are some rules. Chromosomes are like the rules of engagement when it comes to recombining DNA. The way the parents' chromosomes are arranged: that's *sex*. The way those chromosomes tell the baby's DNA to arrange *itself* is sex determination.

Plenty of invertebrates (like cloning wasps) procreate asexually, but it's a lot less common in the vertebrate world: There are only about 70 known species of asexually reproducing vertebrates, including some snakes, some sharks, and a few birds, under the right circumstances.

"The right circumstances" are usually when males are scarce. Weird things happen when you get clones in the mix: the usual rules of engagement do not apply. Nature abhors a virgin birth; it only happens when the population is in danger of dying out.

We're not totally sure how the female's body *knows* if the population is dying out—it has something to do with hormones and the female's interaction with her environment, such as higher temperatures or access to food and water. (It takes a lot of energy to make a new animal all by oneself, and if there's not enough to eat, there's no reason to make another mouth to feed.) Such changes can shift hormonal messages in the female's body and affect her output.

But in the lab, things don't follow nature's rules of engagement. Instead of nature determining whether its genetically necessary to reproduce asexually, researchers are forcing it. Somewhere along the line, in breeding new lab lizards, a lab technician bred one whiptail that had *three* chromosomes. The extra chromosome must have entered the picture when a cloned

THROUGH DARWIN'S EYES

At this time, Darwin wouldn't have known about whiptail lizards' special abilities, but he certainly wrestled with the notion of asexual reproduction, or parthenogenesis, in "higher animals." He knew enough to know that the phenomenon would inform his theory if he could understand the underlying mechanisms at play. He knew that sexual reproduction had benefits, and those benefits were at the crux of his idea of "fitness"—a word that even he didn't like at the time because it felt too prescriptive, too qualitative. In one publication, he breaks down his observations of asexual reproduction from plants to insects, trying to work out what all the organisms have in common,

what it is about the process that, for those organisms, works. He finishes: "we may conclude that the difference between sexual and asexual generation is not nearly so great as at first appears; the chief difference being that an ovule [egg] cannot continue to live and to be fully developed unless it unites with the male element...We are therefore naturally led to inquire what the final cause can be of the necessity in ordinary generation for the concourse of the two sexual elements." In other words, Darwin was pretty unconvinced that asexual reproduction could even occur. If only he could see that whiptail lizard today.

SEX AND GENDER

Scientifically speaking, *gender* and *sex* are not interchangeable. Gender is a human idea that refers to the way a person feels on the inside, or the physical sex with which they most identify. (This is also different from sexual preference, which refers to the type of person you want to have sex with, which is even more complicated still.) Humans feel awkward saying the word *sex*, so they'll often say *gender* when they mean *sex*, as in when an expecting parent or parents does a "gender reveal." Technically, announcing "boy" or "girl" is really a "sex reveal." And because that knowledge almost never comes from a blood test looking at the fetus's chromosomes, it's really a "penis or vagina reveal."

Plenty of invertebrates (like cloning wasps) procreate asexually, but it's a lot less common in the vertebrate world: There are only about 70 known species of asexually reproducing vertebrates, including some snakes, some sharks, and a few birds, under the right circumstances.

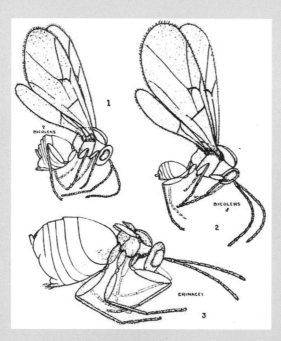

Cloning wasps are known to reproduce asexually; that is, when offspring arise from a single organism and inherit the genes of that parent only.

female mated with a male that was himself the product of a female virgin birth, so the way their unusual chromosomes combined allowed one to stick around. Then, in 1967, a researcher (named Neaves) found a whiptail with *four* chromosomes. He recorded it and named it after himself, but the scientific community largely ignored it for 50 years.

Then, in 2014, a group of American researchers decided to try to bring the whiptail lizard into the age of the genome by sequencing its genome, to get a window into how it got all those chromosomes and what it might be doing with them.

In doing so, they were able to piece together the approximate parentage of the original Neaves whiptail, enough that they thought they could breed another. They did so, by mating one whiptail with three chromosomes and another with two. Eureka. It worked.

Even crazier is that the new species was viable. And prolific. For multiple generations. That original lizard's offspring gave way to more offspring, which gave way to more.

So is this new species actually a species? Or is it a hybrid? Or is it a clone? Neaves researchers advocate calling the new species of lizard a "hybrid clone" to account for its viability as a hybrid, but also for the fact that once it became a hybrid, it reproduced more like a clone.

The terminology might not matter as much in the real world as it does to scientists in the lab, but once these words get out, they have a mind of their own. Just as long as the lizards don't get out of the lab: the marbled crayfish started out the same way, an asexual grand-mommy turned accidental hybrid clone. In 2018, a fisherman released a lineage of the buggers into local waterways, and it started taking over Europe and North Africa, decimating native populations of aquatic life as it went.

LESSON OF THE WHIPTAIL LIZARD: Asexual reproduction can occur. And it can be viable. And even genetically diverse.

TERMS DEFINED: Chromosome

HOATZIN

(Opisthocomus hoazin)

STUCK IN THAT AWKWARD PHASE

The ugly duckling that stayed ugly, the hoatzin is a useful if misleading introduction to the origin of birds.

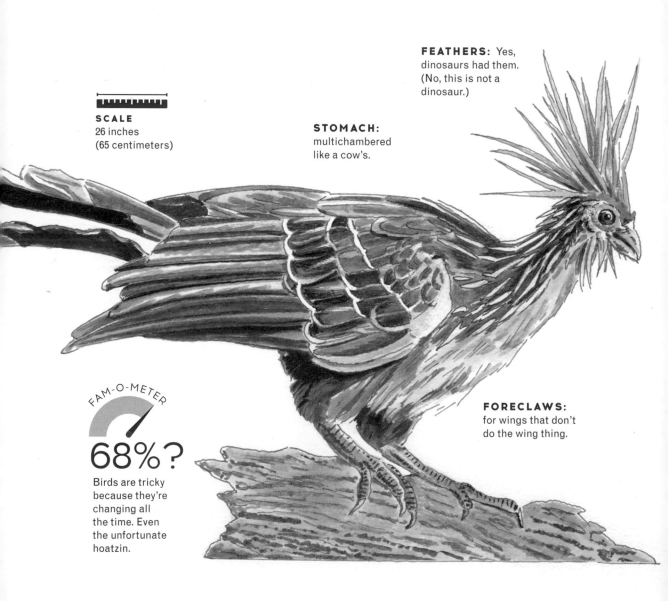

SCALE
26 inches
(65 centimeters)

STOMACH: multichambered like a cow's.

FEATHERS: Yes, dinosaurs had them. (No, this is not a dinosaur.)

FORECLAWS: for wings that don't do the wing thing.

FAM-O-METER

68%?
Birds are tricky because they're changing all the time. Even the unfortunate hoatzin.

BEASTLY BREAKDOWN

FEATHERS

We humans have long associated feathers with flight. By extension, we long assumed that flightless birds had lost the ability to fly from disuse, as if to shame them for their cowardice. Think of the ostrich with its head in the sand, or our meaning when we call someone a dodo or a chicken.

We now understand that flightlessness was probably the birds' original state of being, and one way by which we've come to this understanding is via the origin of feathers. Fossil evidence showed feathers on flightless prehistoric proto-birds. Feathers were just another body covering, connected to the same genes as scales and hairs.

FORECLAWS

The hoatzin may not be a great flier, but what it lacks in grace, it makes up for in claws. Hatchling hoatzin have foreclaws on their wings, just like a bat. They disappear by adulthood, but as hatchlings, they actually have a use for it: adults build nests out over water, and naked chicks jump out into the water, swim back to shore, and claw their way back up the tree.

Thus far, it's unclear if the hoatzin retained the forelimb claws of its 66-million-year-old ancestor, or if, like flightlessness, it lost them at one point and got them back. Either way, I like to think of the hoatzin as the *Freaky Friday* bird, who swapped evolutionary trajectories with its more graceful near cousins to stay in that awkward phase indefinitely.

SOUNDS

The hoatzin barks and caws in guttural tones. Highly diversified songbirds they are not.

STOMACHS

The hoatzin is the only bird in the world that eats nothing but leaves, which, compared to seeds and fruit, aren't very nutritious and are hard to digest. Like other leaf eaters and ruminants (see cows), the hoatzin has evolved a multichambered digestive tract with lots of little "stomachs," where the leaves can sit for a while and be digested by friendly bacteria.

Like other leaf eaters, it belches up methane, the same gas cow farts are made of. And this is why some folks call it "stinkbird."

METABOLISM

Does this mean that long before birds and mammals diverged, their ancient common ancestor had a multicompartmented stomach? No. For one thing, their stomachs' exact structures differ slightly. For another, a 2015 study comparing hooved ruminants, leaf-eating monkeys, and hoatzins showed that all of them had evolved the ability to digest leaves, but in three different ways. All three had evolved a similar set of genes, but the subtle differences among them showed that it had happened at three different times. Like complicated eyes, here we have another example of convergent evolution. It turns out that dealing with toxins, like sighting prey or not being prey, is another one of those key evolutionary drivers.

WHEN IN (PRE)HISTORY/ WHERE ON THE RIVER

250   **MILLION YEARS AGO 65**   **0**

MESOZOIC	CENOZOIC

...Around 219 million years ago, the lineage that would become early birds and dinosaurs split from the lineage that would become crocodilians.

HOATZIN

...This next part is still a little hazy. (See chicken, page 193.)

Let's start with what everyone really wants to know: whether birds came from dinosaurs.

First, let's get some things straight about the term *dinosaur*. We think of dinosaurs as any large, scary-looking animal that lived before cave people and didn't have hair. Actual dinosaurs, in fact, emerged after ancient reptiles had split from ancient mammals. *Dimetrodon*, that huge low-rider with a ridged sail on its back? It pre-dated dinosaurs by over 40 million years and was actually a synapsid, the same lineage that eventually became mammals.

Next to synapsids, the other thing going on back then was diapsids. The two categories refer to a classification relying on skull patterns. Dinosaurs were one of the diapsids group, as was a lineage that would become modern crocodilians and birds. (Closely related were proto-turtles, whose skulls were a little different, but they can still be in the club.) There's an actual name for this club, because scientists like to give names to things: archosaurs.

Then, 66 million years ago, the dinosaurs and most other diapsids of the time went extinct. Thanks to the so-called "K-Pg" event, which wiped out over 70% of what was living at the time, especially the largest animals, and especially the dinosaurs (see page 169).

But for proto-birds that survived, it cleared the way for a revolution, as it did for many of the humble animal groups that survived K-T. Around 65 million years ago, a massive evolutionary burst occurred—a bird big bang, if you will. A tremendous number of new species with vastly diverse new traits appeared in the bird lineage in a crazy short period of time. To our knowledge, it's the biggest and fastest evolutionary boom ever to take place among animals. The resultant birds are collectively considered neoaves, or "new birds."

The short version to the answer, then, is no, birds are not dinosaurs, in the same way that humans are not chimpanzees. Proto-birds lived alongside dinosaurs, but they survived and dinosaurs did not. However, *T. rex* and proto-birds were basically cousins, which is pretty cool. This means that some modern bird is most closely related to the dinosaurs. And you might be beguiled

Comparative anatomy: the left wings of (I) an extinct *Archaeopteryx*, (II) a hoatzin nestling, and (III) some kind of pigeon. From the influential evolutionary text *The Origin of Birds*, by artist and amateur paleontologist Gerhard Heilmann.

into believing that the hoatzin—clumsy, stinky, and crudely clawed—is the best contender.

A 2014 bird big bang study was groundbreaking, but it still couldn't sort out some early changes in bird evolution. And it still couldn't sort out the hoatzin. One of the 40 papers that still appear online has the hoatzin perched awkwardly and alone on the end of its own branch on the neoave tree.

But closer genomic analysis suggests that while hoatzins split from the rest of the neoaves just a few thousand years after the initial big bird bang, hoatzins do seem to have a common ancestor with cranes, plovers, and other leggy birds that also spend most of their time at the water's edge, albeit much more gracefully.

LESSON OF THE HOATZIN: Birds have a lot of genetic material to choose from, which accounts for their extreme diversity. They can even call back long-lost traits, should they become useful again.

TERMS DEFINED: Archosaur

AFRICAN MALARIA MOSQUITO

(Anopheles gambiae)

WHY, THOUGH?

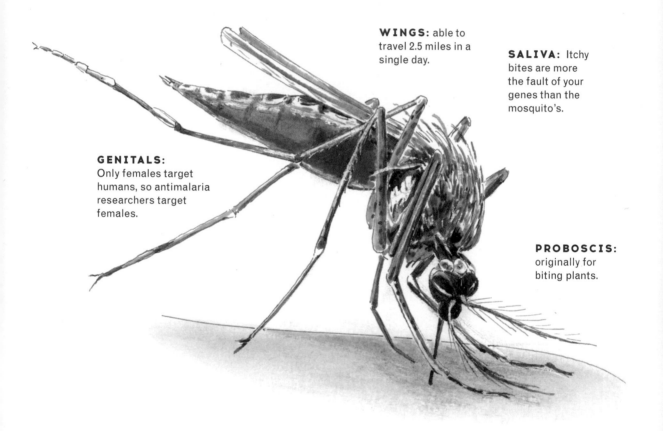

WINGS: able to travel 2.5 miles in a single day.

SALIVA: Itchy bites are more the fault of your genes than the mosquito's.

GENITALS: Only females target humans, so antimalaria researchers target females.

PROBOSCIS: originally for biting plants.

SCALE
0.17 inches (4.4 millimeters)

FAM-O-METER

60–68%

This exact comparison has only been partially sequenced, targeted specifically for two malaria-related gene areas. It becomes harder to parse the difference if the mosquito in question is full of your blood; however, that percentage does go up—which is a joke, but it's also not. Mosquitoes, the microorganisms they carry, and humans have evolved and are evolving together.

Be suspicious whenever you hear someone attempting to answer the question "Why?" with science. *"How?"* Sure. *How* asks for facts. *Why* leads you astray. But that's really all most of us want to know when it comes to animals like the mosquito, which, for humans, range from irritating to deadly. Why, evolution? Why mosquitoes?

WHEN IN (PRE)HISTORY/ WHERE ON THE RIVER

There are about 3,500 species of mosquitoes still in existence on Earth today. The one whose genome has gotten the most attention is the one that spreads the most malaria, for obvious reasons. It is, to genomic specialists, their constituents, and their funders, a matter of life and death.

This mosquito's genus, *Anopheles*, actually has shown up in the fossil record; unlike many of the millions of insect lineages on Earth past and present, a lot of ancient mosquitoes actually have been found in amber. (Although it wasn't until 2017 that one actually turned up with enough blood in it for scientists to start talking about a Jurassic Park scenario in earnest.) The earliest *Anopheles* fossil was a specimen trapped in amber that dated back to the Cretaceous era, about 145 million years ago. So, for a long time, that's when scientists said the genus branched out on its own. But in 2017, a group of American and British researchers cross-referenced mosquito genomes to find that the *Anopheles* lineage may have split from its cousins more like 226 million years ago.

A 2015 study suggests that *Anopheles* split from the fruit fly (*Drosophila melanogaster*, page 137) just 250 million years ago.

MALARIA

According to a 2018 genomic study of malaria parasites, it was just 10 million years ago that the parasites themselves adapted some specialized genes that they still carry today. Transposons indicated that around that time their genomes adapted to incorporate amino acids specific to a particular type of host: birds and mammals, including modern humans.

The *Anopheles* genus itself seems to have continued dramatic genetic splits through around 64 million years ago, since which point most of today's known species have existed. But the traits within them, especially the ones that give them their special symbiotic relationship with malaria, show genetic markers that have arisen much more recently. The only good news is that human intervention can affect that evolution, too.

BEASTLY BREAKDOWN

EXOSKELETON

Invertebrates wear their skeletons on the outside, made almost entirely of a material called chitin. It's kind of like the insect version of keratin, which makes up human hair and nails.

SEX DIFFERENCES

Only female mosquitoes bite, and they can't lay their eggs unless they've had a blood meal. So one clear way to attack malaria mosquitoes, researchers thought, would be to target females. In 2016, Italian anti-malaria researchers utilized this knowledge to create a generation of females that only hatched females, but that would also make those females infertile using the gene editing system **CRISPR**. In 2018, researchers began buzzing about a new CRISPR technique that could produce mosquitoes that are resistant to the malaria parasite in the first place.

WINGS

What's interesting about the *Anopheles* and related mosquitoes is how far their tiny wings can carry them: up to 2.5 miles in a day. In the wild, females have a life cycle of about two weeks. The parasite that causes malaria is contagious for the entire time, ready to start its life cycle anew in whatever host the mosquito chooses to bite. The farther a mosquito's wings can carry it, the more potential victims it can unwittingly infect.

PROBOSCIS

All mosquitoes must eat plant nectars and juices to survive. This diet probably has something to do with why mosquitoes are found caught in amber more often than less sugar-happy insects: tree sap is sweet but a little less forgiving than

CRISPR gene editing is a groundbreaking technology developed over the past decade by numerous teams of geneticists. In a very small nutshell, it uses micro-biotechnology to trick RNA molecules into allowing scientists to cut apart DNA in a way that usually DNA only allows itself to do. Via these artificial means, scientists can swap out sections of DNA, manipulating the characteristics that result in an organism.

nectar. *Anopheles* saliva contains a cocktail of dozens of proteins that help them digest the sugars in plants. Somewhere along the line, that same or a similar cocktail of proteins came to help these mosquitoes digest the blood of animals—which, after all, has some sugar in it as well.

But bear in mind: of the thousands of mosquito species, only a couple hundred drink human blood. They use their proboscis like many of their insect brethren, only they're sipping animal juices instead of plant juices. These mosquitoes, including *Anopheles*, mostly bleed-feed on other animals as well, ideally warm-blooded ones (endotherms) like other mammals (mice, cows, horses, dogs) and many species of bird (including chickens and even barn owls). They've been known to bite reptiles and amphibians as needed, even if it means dodging a sticky tongue.

EXTRA SENSES

Mosquitoes that bite prefer humans as a main course. They track human prey visually, but sniff out their favorites by picking up on heat and humidity signatures and chemical compounds. They can "smell" carbon dioxide, which all mammals exhale with every breath, as well as lactic acid and octenol, both of which exist in mammalian sweat as well. Two recent studies even suggest mosquitoes prefer beer drinkers over drinkers of other beverages, and pregnant women over non-pregnant humans at a rate of two to one. If you've ever asked a mosquito "Why me?" That's why.

SALIVA

The proteins in mosquito saliva trigger allergic reactions in some humans and not others. If you're getting more or worse bites than your friends, it might be your genes' fault.

There are a few specific saliva proteins that correlate to malaria contraction in humans. Over the past decade, researchers from around the world have isolated proteins in mosquito saliva that trigger specific responses in *just some* humans. This combination of human and mosquito genes makes the human in question more likely to contract malaria. Of course, you're also more likely to contract malaria if you're being bitten by more malaria-carrying mosquitoes.

Objectively, the genetic markers associated with salivary glands and proteins serve as evolutionary markers, and help researchers to understand which malaria-linked saliva is related to which. This then helps to trace mosquito lineages that are more likely to carry malaria, and in which genes.

Returning to the question of why, it's important to remember that it's just chemistry. Mosquitoes didn't evolve proteins specifically for biting mammals. Mosquitoes don't benefit from passing malaria on to humans. Once upon a time, mosquitoes coevolved with plants and became able to digest sugars. Later, mosquitoes and vertebrates coevolved to drink blood, have blood, react to the proteins involved in drinking blood–or not. The parasite that causes malaria coevolved with mosquitoes and vertebrates all

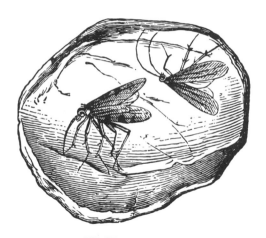

THROUGH DARWIN'S EYES

In 1832, when Darwin wrote a letter home to his beloved Caroline from Rio de Janeiro, it was still decades until science would make the connection between mosquitoes and malaria. Writing only briefly of the death of two of his companions on the *Beagle*'s voyage, Darwin referred to their maladies only as "fevers," saying, "how mysterious & how terrible is their power."

along, its chemicals (like proteins, enzymes) working in concert with ours, with mosquitoes', until all parties arrived in the place we are today. The only "why" that can matter to us is why genetic research needs funding. "Why" is a human question. And the answer to that question provides human solutions.

LESSON OF THE MOSQUITO: Evolution has no endgame in mind. Physical traits can change their function. Sometimes animals evolve together, and it's good for one or the other, sometimes bad, ultimately neither.

TERMS DEFINED: CRISPR

BRAIN: hardwired not to know any better

COAT: what separates the dogs from the wolves (especially coats with spots)

EARS: can hear about 4x better than a human

NOSE: can smell about 14x better than a human

FAM-O-METER

85%

This percentage is higher than mice and lower than cats, despite the fact that we share a more recent lineage with mice than with either. Why is that? Remember: genetic overlap can come from a common ancestor or can develop anew based on shared circumstances. Dogs and humans have shared their circumstances for the entire time that both have taken their modern forms. I'm not saying there would be no dogs without humans and no humans without dogs, but, well, it's nearly impossible to parse what we'd be like without each other.

SCALE
8 inches (20 centimeters) for a Chihuahua to 34 inches (86 centimeters) for a Great Dane

DOMESTIC DOG

(Canis lupus familiaris)

WEIRDO WOLVES WITH HUMANS FOR A PACK

The modern dog exhibits about as much anatomical variation as you can see in a single species of animal, and they have humans to thank. But from mutts to purebreds, they all (probably) started in the same place. Ultimately, there are two parts to the evolution of the dog: BU and AU: before us and after us. Before us, dogs were basically wolves. After us, dogs got less scary and more friendly, and everything beyond that was completely up to us.

WHEN IN (PRE)HISTORY/ WHERE ON THE RIVER

10	MILLION YEARS AGO	0
	CENOZOIC	

...Within the last 2 million years, the wolf lineage split from their common ancestors with coyotes, then a little later their common ancestor with jackals and dingoes.

DOMESTIC DOG

...Around 10 million years ago, wolves split from their common ancestor with the lineage that would include foxes, which is interesting, because even modern dogs can successfully breed with foxes, and foxes can be domesticated.

...Around 1 million years ago, there was wolf. Prehistoric wolf species came and went, and in the modern era, *Canis lupus* lived throughout North America, Europe, and some of North Africa.

The moment when the wolf took its first step toward dogliness is no more specific than any other moment in evolutionary history. Humans and wolves both evolved to be social apex predators who lived and hunted in packs. For both, their existence was all about their pack, the source and purpose of their survival. Somewhere along the line, some wolf or some human decided to reach across the aisle and help out some individual from the other group. Knowing what we know about wolf pack dynamics, the rogue was probably a wolf. When a young wolf grows up, they usually leave the pack and go in search of a mate to start their own pack. This is what we think of as a lone wolf, but they'd prefer not to stay that way; they've just outgrown their pack and are transitioning toward the next. If one such wolf came across a pack of humans before they found a new family, it may have been easier to linger around their perimeter in hopes of stealing a scrap—especially in the thick of an ice age. If, while lingering, the wolf's keen hearing or sense of smell alerted the humans to potential danger, well now we're really onto something.

Eventually, this type of cross-species partnership became a symbiosis neither group could live without. The scientific community continues to argue over the specific timing and genetics of dog domestication, but a 2016 genetic timing study seems to have found a number most can agree on: 40,000 years ago, give or take. By that time, the dogs that hung out with humans were mating among themselves more often than with wild wolves. And so it started.

As humans progressed, dogs progressed. Or rather, they progressed together.

The remains of a dog that died around 14,300 years ago appear to have been ceremoniously buried with her human companions, according to a 2016 study.

Twelve thousand years ago, as ancient Mesopotamians settled around fields of wild edible grasses like barley, large, proto-mastiff-type dogs settled with them. These dogs would have guarded against dangers to the people and their food source. Grazing animals like deer might otherwise have eaten away the wheat and barley grasses before humans figured out how to make it edible. Luckily, grazing animals are the canid's favorite prey.

In the millennia that followed, as ancient people mastered the cultivation of grains, early dogs would have caught on that humans were happy with any and all thwarting of threats to their families or food. (Cats helped here, too, but we'll get to that later.) The extra warm bodies didn't hurt either. By 9,000 years ago, people in colder climates like Alaska and Siberia had harnessed the power of the sled dog, the pack that serves as both transportation and cover, in a pinch.

Dire Wolves Were a Real Thing:
Around 11,700 years ago, at the end of the last Ice Age, the dire wolf roamed. A real, actual dire wolf, not a creature from some fantasy novel. It was about four times bigger than a modern bull mastiff, and four times heavier than the largest modern wolf. Studies haven't yet found any dire wolf–specific genes in modern dogs. Though modern dog lineages may have overlapped in time with dire wolves, they may not have overlapped in space. Thus far, all traces of it died 10,000 years ago.

By 5,000 years ago, ancient peoples had mastered the cultivation of grains. Do humans owe the very rise of civilization to our relationship with dogs? Honestly? Maybe. It was probably around this time that humans started actively breeding dogs for certain traits, kicking off a process of artificial selection that continues today. Egyptians cultivated barley and wheat and the early ancestors of today's pharaoh hounds. In Asia, humans cultivated rice and pariah dogs. In Mesoamerica, the Mayans raised corn and the early ancestors of the Mexican hairless. By 3,000 years ago, breeds began to look the way we'd know them today: the Mediterranean Maltese, the Chinese Shiba Inu, the Bankhar (Tibetan Mastiff). They'd become so domesticated they could be trained to guard and herd the animals their ancestors would have hunted: the Bankhar's specialty is killing wolves.

All dogs, of every breed—every last domestic dog on Earth today—is considered the same species. The variation among these breeds is a testament to the mysteries and power of the genome and to its depth, really, if certain areas can change nearly everything we can observe in an animal, while barely changing a thing in the invisible majority of genetic information. Even though humans have spent millennia carrying out this selection process with incredibly specific intentionality, we hadn't really been sure what we were changing from a genetic perspective. In 2017, we got a little closer to understanding. After working for 20 years to collect and compare the genomes of 1,346 dogs from 161 breeds, a group of American genomicists published a canine tree of life. It outlined the probable when and where origins of almost half of all known breeds.

In a 2018 study, researchers compared DNA from 5,000-year-old remains to DNA from modern wolves and DNA from village dogs from various cities around the world—village dogs being mutts that live among people but with no one in particular. The study ran the three types of DNA through algorithms that would highlight differences, which would in turn highlight genetic changes that had come about with domestication (defined as depending on humans for food). This stage of domestication appears just 300 years ago.

BEASTLY BREAKDOWN

BRAIN

The most important change that made dogs domestic was in their temperament. Gentle, helpful dogs joined the human fold to feed and breed alongside us. Today, dogs are so attuned to human habits, expressions, voices, and smells that they can sense details about us even before we can—our moods, our health, our intentions. While it's difficult to track the genetic origins of friendliness, one can look for patterns in inheritance to see what other traits are associated with friendliness.

In fact, the first person to document a pattern of traits associated with friendliness was Charles Darwin. Nine years after first publishing *Origin*, he still hadn't cracked the code of *how* animals inherited and passed along traits. He'd spent the decade studying domestic

WHO'S AFRAID OF THE BIG BAD WOLF?

Wolves almost definitely never hunted humans. Humans and wolves are both predators, and predators evolve to get better at hunting the prey animals they're best at hunting, just as their prey animals evolve to get best at not getting eaten (or at surviving as a group, even when some of them are eaten). Predators evolve to find niches and ways of coexisting with other predators. For instance, wolves eat meat, and specifically hooved animals if they can get them. Subspecies that prey on bison have evolved to be stronger, whereas subspecies that go after antelope are faster and more agile. And in habitats where proto-wolves, proto-bears, and proto-humans evolved in the same area, it's no coincidence that primates and bears evolved to eat other things if need be.

So consider fairy tales about wolves eating grandmothers, Riding Hood, and so forth as nothing more than anti-wolf propaganda. Today, a few records of predation on humans have begun to spring up in areas where drastic habitat changes have simultaneously put apex predators in desperate situations *and* close proximity to humans—as with some wolves and wildcats in India and Africa, and polar bears in Alaska. But these animals have also been known to cannibalize their own kind when no other options present themselves. Which, unless you've evolved to have that be a thing, like the flour beetle, is really not ideal.

ABS, Always Be Skeptical: This 1910 illustration depicts a gray wolf snarling over its kill. Accurate: That it would snarl to defend its kill. Less accurate: That it would have killed a bird. Wolves usually hunt as a pack and go after large, hooved prey.

animals especially carefully, and he had noticed that they had some things in common. Compared to their wild relatives, tame dogs had smaller teeth and jaws, floppier ears, a more curved tail, and white spots that would show up in the coloring. Even stranger, he noticed that similar changes occurred in other domestic mammals—cattle, goats, pigs, cats, rats, rabbits, mink, donkeys, camels—and even some species of domesticated bird and fish. We now know that the phenomenon can occur in most mammals, even in apes and humans.

Darwin dubbed the phenomenon "domestication syndrome." We also now know that the syndrome also includes internal changes: a reduction in brain size, especially in the regions that control executive function and fear response; changes in several groups of neurotransmitter, changes in hormone levels associated with adrenaline; menstrual cycles that become more frequent and fall out of sync with the seasons; and prolonged periods of juvenile (baby-like) behavior.

Darwin never did figure out what was causing the groups of changes, because it was, of course, genes. And we now understand that genes, chunks of DNA, can be connected to one another on the inside of an animal even if they wind up affecting seemingly unrelated traits on the outside. Many researchers after him continued to examine the subject and found more patterns. The traits really did seem most associated with docility and lack of fear. If a breeder kept foxes for generations and bred for the floppy ears or tail alone, they might still have wild-acting foxes several generations later. Meanwhile, if they bred for friendliness, the whole family of foxes would have all traits of the syndrome in just a few generations. This meant that domestication syndrome wasn't about

actually living in captivity. In fact, families of animals with domestication syndrome would still have smaller brains 40 generations after being released back into the wild. The key to the syndrome was in a set of crucial genes that, once inherited, couldn't be uninherited and couldn't easily be overridden.

A major milestone in this subject came in 2014, when an international group of researchers officially outlined something generally called the neural crest hypothesis. The hypothesis suggests that domestication syndrome all comes down to a set of embryonic stem cells—cells that are floating around inside the animal before it's born, when it's still an embryo developing in its mother's egg and its stem cells haven't quite decided what they're going to become. Stem cells have all the DNA they're going to inherit; they just haven't yet put it all into action. That happens as the embryo develops.

There's a certain moment in development when a wave of stem cells float down the embryo's little head and all start to do their thing. These stem cells are collectively called neural crest cells. Most of them wind up kicking off brain functions, like brain growth and development of adrenal systems, but some of them have other developmental duties, like building cartilage in the ears and tail, or starting the growth of teeth.

Finally, the pieces of one of the oldest conundrums in evolutionary biology seems to be coming together. There are still a lot of questions. The movement of stem cells is something that happens after an animal has inherited all the DNA they're going to inherit. So how does something going wrong at this point then become inheritable? (This is one of those epigenetics questions that are so controversial right now and represent one of the next big frontiers of evolutionary and genetic science.) But at least we're a little closer to unraveling the very complicated story of how the cunning wolf evolved—coevolved with us—into the sweet and simple dog.

COAT

As a trait affected by neural crest development, the domestic dog's coat started showing white spots where its wolf progenitors only had solid bands of color.

But far before we understood that, humans started breeding dogs for changes to their color and texture depending on what they needed them for. Poodles, for instance, originally bred to help hunt ducks, have wiry water-resistant hair that doesn't shed with the season. Meanwhile, several breeds of greyhound, which were first bred to hunt rabbits in the deserts of the Middle East and North Africa, have such short fur their bellies are practically bald.

JAW AND TEETH

Another trait affected by neural crest development, the domestic dog's jaw got shorter and less powerful and their teeth got smaller.

NOSE

Dogs have retained the wolf's excellent sense of smell, which is one of the traits that has made them so useful to us. With sniffers as many as

ABS, Always Be Skeptical: The artist for Darwin's *On the Expression of Emotions in Man and Animals* took some liberties with this dog's expressive eyebrows.

14 times keener than humans', dogs can help us detect prey we might otherwise miss and alert us much sooner to potential threats to our safety.

It's all thanks to a highly developed nasal membrane that sits at the back of their snouts, all the way up their nose. The membrane is packed with over 300 million olfactory receptor nerves, lined up on hundreds and hundreds of folds, which allows maximum surface area for its size. If you stretched the membrane out, it has a surface area of 393 square feet in some dogs (120 square meters). Humans' nasal membranes are about 65 square feet (20 square meters).

Modern breeds can be trained to recognize and track almost anything: explosive residue, drugs, cadavers, even the specific metal of a landmine buried up to 20 feet (6 meters) underground. Bloodhounds are so named because they can differentiate the scents of individuals from the smell of their blood and can track a human through an area as many as four days after they've been there. They were supposedly first bred in a medieval Belgian monastery. Breeders even found a way for the bloodhound's ears to factor into its job: they act as visual and olfactory blinders, funneling scents from the ground up to the dog's sniffer.

EARS

Keen hearing is one of the primary traits that made wild wolves successful predators and makes dogs useful to us in their role as guards and hunting companions. Popular wisdom holds that wolves can "hear" sounds that are as many as 6 miles (9.5 kilometers) or even 10 miles (16 kilometers) away, but that all depends

Why does my dog roll in stanky stuff? Why? One behavior that's probably a consequence of this super sense is that thing where dogs roll around in smelly rotten dead stuff. Wolves do this, too. The reason why remains unclear: One hypothesis was that they would bring the smell back for the rest of the pack to analyze, but evidence of the actual scent sharing has yet to be recorded. It's possible that, for animals whose sensory lives are far more about the smelled than the seen, overwhelming putrescence is just a sensual feast for them, the equivalent of a bed of luscious silky fabrics that they're just powerless not to revel in.

on the type of sound, does it not? We think they can probably hear each other howl at that distance. We also know that domestic dogs can hear considerably better than us, especially at frequencies over 250 hertz (and all the way up to 26,000 hertz).

Ears are also affected by the movement of neural crest cells, which are floppier in domestic dogs than in wolves. Beyond that, pricked versus folded, triangular versus rounded—those details were choices humans made. Or they were traits that wound up being genetically tied to other choices that we made.

Mammalia Pl. 4

Falkland Island wolf, by Captain FitzRoy of the *HMS Beagle*

FEET

Prehistoric canids evolved a strong webbing between their toes that helps modern wolves walk and run long distances over changing terrain. It even acts as a sort of snowshoe, distributing their weight such that they can remain more agile on snow drifts while their hooved prey sinks into them. Humans exaggerated this quality in certain fishing breeds like Labradors and Portuguese water dogs, whose extra-webbed feet help them swim without tiring.

EYES

Wolves have excellent eyesight, a crucial trait for a predator. This would still be true for dogs, except that the practice of breeding for purity has kept at least eleven different genetically inherited eye diseases in the gene pools of at least 50 breeds, including glaucoma; hereditary

cataracts; rod, cone, and retinal dysplasia and degeneration; and one condition simply called collie eye anomaly. A 2014 study traced the variety of diseases to as many as 29 different gene mutations.

A healthy dog's eyesight should be good even at night, however. Wolves hunt in low light and darkness, so they have a reflective lens called the tapetum at the back of their eyeballs, the one that catches headlights in the dark and makes the eyes of dogs and cats seem to glow green.

Dogs' sight is not great when it comes to color, though. While it's not true that they see only in black and white, their spectrum is simpler than ours: they can see yellows and blues, but instead of reds they get browns, and instead of greens they get grays. Keep that in mind the next time your little buddy loses that fire-engine-red frisbee on a field of green grass or a tennis ball in the snow, and you wonder why they've lost their mind. They're just sort of red/green colorblind.

CIRCADIAN RHYTHM

Beyond the neural crest hypothesis, the other interesting genetic difference between wild and domestic dogs is in a gene that seems to be related to circadian rhythm and sleep. Wolves can be partially nocturnal animals, but domesticated dogs keep the same hours as most humans. If you're a (human) night owl, you may even notice that your dog sleeps through the night, even if you're up and about, and is usually raring to go some time in the morning, even if you're not.

LESSON OF THE DOG: Dogs are man's best friend, man's best genetically modified organism, and man's best walking, shaking, pooing example of real time (co)evolution by our own hands in our own homes.

TERMS DEFINED: Neural crest hypothesis

THROUGH DARWIN'S EYES

We can see in Darwin's observations of the dog that he attempted to marry the narratives of its domestic and wild side. Though he didn't know the mechanisms behind his domestication syndrome theory or the related neural crest hypothesis, the implication here certainly does seem to be that the difference between wolves and dogs is that dogs are just a few crayons short of a box. Dogs, when they wish to go to sleep on a carpet or other hard surface, generally turn round and round and scratch the ground with their forepaws in a senseless manner, as if they intended to trample down the grass and scoop out a hollow, as no doubt their wild parents did, when they lived on open grassy plains or in the woods. Jackals, fennecs, and other allied animals in the Zoological Gardens treat their straw in this manner; but it is a rather odd circumstance that the keepers, after observing for some months, have never seen the wolves thus behave. A semi-idiotic dog—and an animal in this condition would be particularly liable to follow a senseless habit—was observed by a friend to turn completely round on a carpet thirteen times before going to sleep.

Sad as it is to say, a lot of dogs' stranger habits seem to be odd evolutionary short-circuits. But there we go again with our moral judgments and our socionormative criteria. Humans have evolved plenty of habits that don't make sense, either.

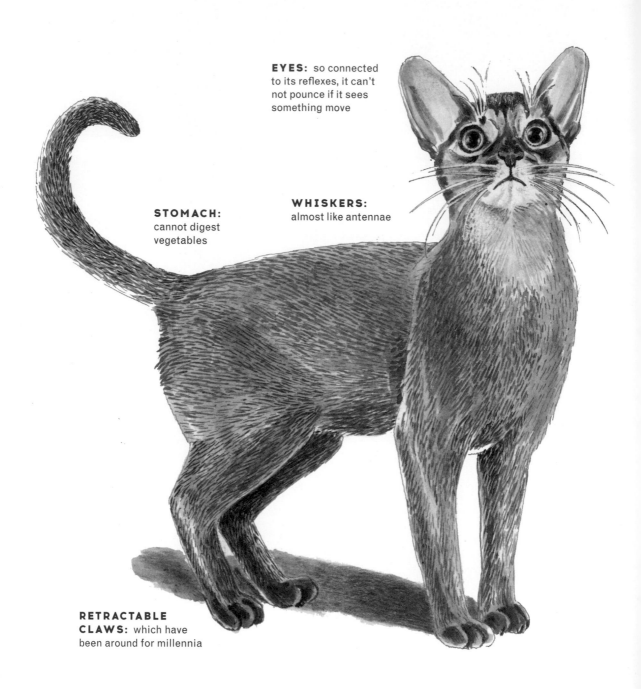

EYES: so connected to its reflexes, it can't not pounce if it sees something move

STOMACH: cannot digest vegetables

WHISKERS: almost like antennae

RETRACTABLE CLAWS: which have been around for millennia

FAM-O-METER

90%

Most of the genes inside of you also show up in your cat. These genes are homologs, meaning they come from the last common ancestor we shared with cats, about 92 million years ago. We're more closely related to cats than dogs, much to the dismay of dog people...and cats. Because if your cat acts like it doesn't like you, that's not in your head. That's in your cat's genes.

SCALE
Average 2 feet (60 centimeters), 10 pounds (4.5 kilograms), or exactly the size of a common house cat.

CAT

(Felis silvestris catus)

LIVING ITS BEST (DOUBLE) LIFE

Dog lovers can easily see the wolf remnants in their furry friends, but the cat's story is at once more complicated and much more simple. Cats didn't evolve from lions or even saber-toothed tigers, but that doesn't mean they don't have a wild streak. What little domestication cats have undergone hasn't done a lot to change them. And it's mostly been on their terms.

WHEN IN (PRE)HISTORY/ WHERE ON THE RIVER

Scientific names tend to reflect the namer more than the named, but in this case, the two are utterly intertwined. For many years, the Linnaean Latin name for the domestic cat was *Felis catus*. "Cat" comes from the Middle/Old English *catte*, which came from the Proto-Germanic *kattuz*, which came from the actual Latin *catta*, which might have come from the indigenous North African *kaddîska* (wildcat), or the Nubian *kadīs*, which probably came from the Arabic *qitta* or *kitt*—one of four words for cat in the Islamic tradition, none of which make a distinction between a domestic cat and wildcat. The Chinese character for *cat* looks like a picture of a cat and is pronounced "mao," which is about as close as you get to an animal telling you its name. So as long as we've had language, we've had cat.

But what's the scientific story? In 2001, an American zoology student set out to travel the world on his motorcycle and collect the largest, most diverse sample of domestic cat DNA to date. On every continent save Antarctica, he swabbed house cats and strays and snipped skin from roadkill, with no cats killed for his curiosity. After bringing home 979 samples that needed processing and analysis, his pet project took him another decade to complete. But his findings were overwhelming: every tom, kit, and tabby had descended from a single species of small wildcat, *Felis silvestris* ("cat of the woods"), which still exists today. Specifically, they all came from *Felis silvestris lybica*, a subspecies with the sandy coloring and "mackerel" pattern stripes down its ribs. Crazier still, their mitochondrial DNA (mDNA, which is passed down through mothers) all fell into five distinct patterns. Five mDNA patterns = five mothers. According to their genes, all domestic cats on Earth descended from just five mothers, tens of thousands of years ago.

10,000-12,000 YEARS AGO

...cats emerged in the Fertile Crescent, the area in and around modern-day Iraq where modern civilization began, which suggests that it takes a truly civilized society to domesticate the cat. Coincidence? Not at all. Both cat domestication and civilization emerged largely because humans started domesticating livestock (See cows, page 95). Storing grain for livestock means rodents, and cats keep those pests away. Archaeologically speaking, the first signs of domesticated cats occur through the ages in paintings, sculptures, and even graveyards...

9,500 YEARS AGO

...on the island of Cyprus, cats wound up in human/pet cemeteries even though the island had no native feline population.

5,000 YEARS AGO

...in ancient Egypt humans buried cat mummies alongside human ones.

1,500 YEARS AGO

...cats appeared in the Middle East, where the Muslim world's already millennia-deep love of cats was indelibly written into tradition via dozens of cat shout-outs in the Qur'an. Prophet Muhammad apparently had a cat that he loved so much that he once cut off the sleeve of his coat where it slept rather than wake it up. In the 13th century, the monarch Mamluk sultan al-Zahir Baybars set up a "cats' garden" in Cairo, where townspeople would bring fresh food for the local cats to make sure they were happy and loyal.

So did cat lovers in each of these locales start selectively breeding their very own wildcats? No. A 2017 study out of France analyzed the mitochondrial DNA of 209 ancient cat carcasses and found regional differences, just as Carlos, the American zoology student, and his team had done with his nearly 1,000 modern cats. Their findings suggested at least two different methods of domestication.

In the Fertile Crescent, the Eurasian lineage *F. silvestris lybica* probably self-domesticated: where humans went, they would follow. At first the wildcats ate humans' post-hunt meat scraps. After hunter-gatherers settled down, the wildcats ate vermin, which ate humans' harvest scraps. The genetic spread of Eurasian *F. silvestris lybica* perfectly aligns with the human spread, following the same routes of trade and exploration outward from the Tigris-Euphrates valley.

The genetic Google Map of the North African *F. silvestris lybica*, meanwhile, was separated from the Eurasian catsplosion by thousands of years and the Mediterranean Sea. Human Egyptians probably had more agency in this case, choosing to bring wildcats into their homes to lounge under tables and be celebrated as the goddess Bast.

Slowly, the Eurasian and Egyptian lineages met up and blended, ever becoming less *sylvestris* and more *catus*. But there's a reason researchers have added *sylvestris* back into the name. While they're still working out which traits come from where, it's clear there's still plenty of wild left.

Felis silvestris, as depicted in the *Meyers Lexikon* German encyclopedia, **1897**.

ACCORDING TO DNA ANALYSIS AND GOOD OLD-FASHIONED SKULL COMPARISONS, FELIS SILVESTRIS:	**... Shares an immediate branch of the river with extant wildcats like the leopard cat (not to be confused with an actual leopard), sand cat, and the jungle or swamp cat from the Middle East and nearby.**

... Another branch over are extant wildcats in the *Prionailurus* genus, like the fishing cat of East Asia, which looks like a small ocelot and is obviously really good at fishing, and the long-haired Pallas's cat or *manul* of Central Asia, which looks ridiculous, like a fluffy Persian with a tabby's face and a bad haircut.

... Another branch after that includes the cheetah, the cougar (aka puma, aka mountain lion), and the jaguarundi (a Central and South American cat that almost looks like an otter, with dark brown fur and a long body).

Members of the subfamily Panthera (lions, tigers, jaguars, and leopards) are the most distant relatives of the furry friend in your easy chair.

Cats are most far removed from saber-toothed cats, a catch-all term that refers to a huge group of extinct cats that long ago split from modern cats, all "conical-toothed" cats. Saber-toothed cats, conical-toothed cats, and an in-betweener with both conical and saber-tooth features last shared a common ancestor about 20 million years ago.

BEASTLY BREAKDOWN

EARS

According to genome comparisons, cats at home and in the wild have the broadest range of hearing in the carnivore lineage, allowing them to hear ultrasonic movements like the chewing of tiny rodent teeth.

NOSE

In comparing their genomes to dogs' (page 115), it seems that cats have traded their sense of smell in favor of other traits over time. But cats have a better sense of pheromone detection than most other animals in this part of the genetic river. In the wild and at home, cats are extremely territorial and use a series of glands to spray and rub a pheromone-rich oil on anything they want to mark as their own. They have glands around their butts, paws, lips, and cheeks. So when your cat rubs its face on you affectionately…? Yeah. Territoriality.

THROUGH DARWIN'S EYES

"Why cats should show affection by rubbing so much more than do dogs, though the latter delight in contact with their masters...I cannot say."

WHISKERS

Whiskers are connected to bundles of sensitive nerves and highly reactive muscle tissue that help the whiskered stay more in touch with their surroundings. They sense changes in air current and measure narrow escape routes, even in the dark. For the earliest mammals, little critters in a land ruled by giant reptiles, newly evolved whiskers may have been one of the best advantages over rivals or would-be predators (more under mouse, page 183). In the safety of the home, though, a modern cat will keep its whiskers away from the face when it's relaxed but pull them tight along their face when they're threatened, possibly to keep them from being damaged in a fight. With all those nerve endings, yanked whiskers would feel more like someone tearing out your fingernails than tugging on your mustache.

JAWBONE

Every modern-era *catus* descended from *F. silvestris*. However, in 2016, researchers from multiple disciplines collaborated to analyze those 3,500-year-old Chinese cat bones mentioned before. The ancient cat's DNA was too damaged for analysis, but by examining its jaw and teeth, osteology experts determined that these ancient Chinese cats were more closely related to the jungle cat than to *F. silvestris*. Folks in ancient China had domesticated some wildcats, but over time, as trade advanced and China made more contact with outside populations, ancestral Chinese cats' DNA was eventually subsumed by DNA from *F. silvestris*.

BRAIN

X-rays of the brains of several species of domesticated animal (dogs, cows, chickens) all show that parts of their brain are less developed than their wild counterparts, specifically the stringy parts underneath and in the middle of the brain that control adrenal response (the medulla, hippocampus)—the parts that control fear response and memory. X-rays of domesticated cats show the same shrinkage, but not to the same extent as in other domesticated animals. Again, this is likely because cats haven't been as fully committed to domestication for as long as other domesticated animals. X-ray analysis of those cat mummies in Egypt showed brains as developed as wild *Felis lybica*.

Even weirder, this brain shrinkage seemed to always accompany a few other traits as well, including small white spots in the coat and floppy ears. The root cause of the multifaceted domestication syndrome turned out to be a shortage of certain neural stem cells in the animals' embryos as they developed. Genetic analysis showed that domestic cats lacked these cells—but again, not as much as their other homies in domesticity.

Cats' place in the home, though, still relies on their special ability to find and hunt small things in dark places. As best we can tell, far more of their neurons are devoted to detecting and reacting to movement than human and dog brains do, for instance, giving them far more "frames per second" of processing power. This might be why your cat's reaction to that cat toy seems almost compulsive. Its brain is overwhelmingly hardwired to pounce.

EYES

Precision predators need especially good vision, specifically binocular vision, meaning both eyes face forward in the head and can work in tandem to judge distance before an attack. Felines, canines, primates, and birds of prey all have binocular vision. Their long-headed prey, like rodents, fish, deer, and songbirds, all have monocular vision, where they can see farther behind them but the two visual fields never cross in front. But even a cats' specialized eyes have limits: they can't properly focus on anything closer than about 11 inches (30 centimeters) to their face. Luckily, that's (literally) where whiskers come in.

The vertical slit-like pupils of house cats are

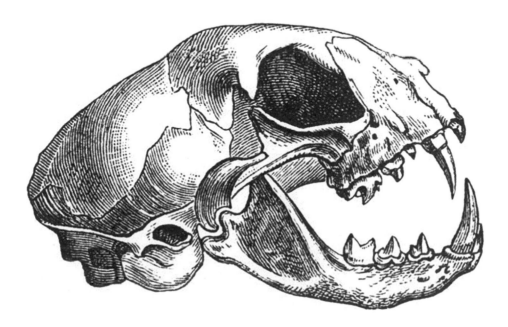

Skull of *Felis catus*. From specimen collector Hubert Ludwig's book *Schul-Naturgeschichte* (School of Natural History), 1891.

another optimal adaptation for gauging distance, specifically low to the ground and at night. Consider, for instance, that the nocturnal python has vertical pupils, while the tall, diurnal (day-active) tiger does not.

Under it all, though, cat eyes and human eyes have a lot in common. The genes that control eye development in humans and cats are homologous, meaning they exist in both animals because they originated in a common ancestor. A primary reason the cat genome project started in the first place was because its subject, a purebred Abyssinian named Cinnamon, came from a long line of cats that suffered from a genetic vision loss called retinitis pigmentosa (RP), which also affects humans. Isolating this genetic locale in Cinnamon has helped RP researchers get closer to a cure.

KIDNEYS

Cats get feline versions of many human diseases, especially diseases associated with old age, including type 2 diabetes, asthma, feline AIDS, two kinds of virus-caused cancers, and kidney disease. Even before Cinnamon's genome was complete, cat-loving geneticists in Missouri isolated a gene that heavily contributes to renal failure in cats that wound up having a homolog in humans. Treatments developed using cats are already beginning to work in humans.

COLORING

In 2007, a group of Swedish researchers took an initial pass at sequencing the genome of Cinnamon. There were errors in that early sequencing attempt, but it was a start. It took almost seven more years to gather funding to try again; according to one insider, they kept losing

The Pampas cat (then *Felis pajeros*, now *Leopardus pajeros*) of South America. By Captain FitzRoy of the *HMS Beagle*.

funders to a dog genome project. But in 2014, a cross-institutional, multinational group of researchers resequenced Cinnamon's genome, along with the genomes of a Russian tuxedo cat named Boris, and a European wildcat whom the researchers named Sylvester. The findings were much more complete, both for their detail and for the comparison possible across three animals with different traits and countries of origin. The animals' commonalities strengthened the universality of the cat genome findings. The origins of their coats, on the other hand, were not so simple.

Felis lybica and her offspring were all one color, which we know by comparing the wildcat's genome to paintings of cats all over the ancient world. In modern cats, variations of hair (and likewise skin) link not to one gene but a combination of many—just as in humans and Neanderthals. Some traits can occur in any breed: calico coloring, for instance, comes from a recessive gene on the X chromosome, so it almost always occurs in female cats. All-black and all-white cats are cases of hyperpigmentation or under-pigmentation in which the mutation overrides all other coloring markers. But coloring patterns that are specific to breeds, like gloving, tuxedos, or silver coats, came about from a random mutation, tied to other mutations, that breeders bred for over time. Some of the coat traits link to other traits like face shape or temperament, while others don't. When a cat is different enough from its peers, it gets a new name. Farmers and breeders, in influencing the course of evolution, or "artificial selection," as Darwin would call it, have known for decades what science is only just starting to spell out.

SEX ORGANS

Though we've cohabited with cats for at least 10,000 years, we've only been actively influencing their sex lives for the last 100—a stark contrast to all other domestic animals whom we've been husbanding since right out of the

gate, so to speak. It's still a difficult thing to do. Female cats can be impregnated by several male cats at once, so "selection" is already convoluted for any cat that spends any time on her own (which all cats, genetics very strongly suggest, want to be doing). Today, as ever, there are far fewer cat breeders than cats breeding, scientifically proving what they say about herding cats.

SENSES

Somewhere between the wild and those five maternal ancestors, *catus* lost the ability to taste sweet. The change could have had any number of causes or more a combination of several—a genetic mutation that made them more susceptible to certain diseases. And probably not unrelated, cats eventually lost the ability to digest or metabolize sweet foods like grain and fruits, and when fruit went out, so did the veggies. The cat's earliest ancestors ate mostly meat (a habit called hypercarnivory, aka your next band name), but with this adaptation, cats became obligate carnivores, or meat eaters by necessity.

CLAWS

Cats have had claws since before the sabertooths split from the cone tooths, but newer research suggests that the retractable claw actually evolved twice: once in the *Felis* line, and separately in *Panthera* in a rare scientific coincidence. Claw retractability allows for quiet walking and therefore stealthy stalking, and it keeps the claws sharp until they're needed, which is to say it's a natural adaptation for any animal with the lifestyles of both lines of cat. Each adapted according to the need, just as both lines of cat ended up as hypercarnivores, but only one lost the ability to eat anything non-meat.

And the claws are still evolving. Cheetahs have evolved to retract their claws only halfway, which gives them the traction they need to catch gazelles and continue being the fastest runners on Earth.

Cat in an Affectionate Frame of Mind, by painter Thomas Waterman Wood. Commissioned by Darwin for his book *The Expression of Emotions in Man and Animals*, 1872.

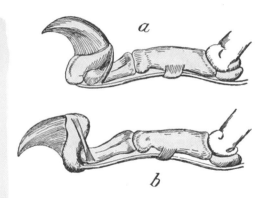

THROUGH DARWIN'S EYES

In 1872, Darwin, then in his 60s, was living a much less exotic life in England and seemed to be spending a lot of time watching kittens nurse at their mother's belly, because he referenced them a *lot*. It was one year after the publication of *The Descent of Man*, his famous/infamous overview of human evolution, and he published his third work, *The Expression of the Emotions in Man and Animals*. He had written it as part of *Descent*, but had pulled it at the last minute, worried that the public was already poised to tear his new work to pieces and that the bit about emotions might read as unscientific. He proved right on both counts. The latter work got a few things right and a few things wrong, including in an entire section titled "On the Special Expressions of Cats." In it, he correctly identifies nursing as the reason cats do that kneading thing with their paws, but he's wrong about it being the reason a cat "rubs her cheeks and flanks against her master or mistress...against the legs of chairs or tables, or against door-posts." He correctly notices that cats clean themselves with their tongues, wondering: "Their tongues seem less well fitted for the work than the longer and more flexible tongues of dogs." He doesn't seem to have seen the tongue under a microscope, or else he would have noticed its tiny, specialized coat-cleaning hooks.

METABOLISM

As obligate meat eaters that live among humans, domestic cats have also adapted to metabolize fats without developing the kind of coronary heart disease or plaque-y arteries humans get. Wildcats don't have the same adaptation, but they also only eat game, and only what they catch themselves. This adaptation protects cat hearts, but it doesn't protect them from gaining weight. Evolution has also adapted domestic cats to have longer intestines, which means more room for food and more nutrient absorption. Combine more room in the gut, no consequences for overeating, and active food-seeking behaviors both in and out of the home, and you get fat cats. And say what you will about the scientific credibility of body mass index (BMI), but it's no wonder that domestic cats can have BMI numbers up to 75 and counting, a whopping 65 points above the average *F. lybica*.

LESSON OF THE CAT: Right down to the genomic level, this is what it looks like when an animal has coevolved with humans but isn't being all codependent about it.

TERMS DEFINED: Mitochondrial DNA, exponential growth

PART 2:
SONG OF MYSELF

Some basic tenets of evolution are all well and good, but what does this have to do with us? Humans don't have long necks or color-changing wings, but working within us are those same pressures from various types of selection, the same mechanisms of protein making, transposon-swapping genes that inform one another in unexpected ways. We, too, it turns out, have coevolved. We, too, are hybrids. The evolution of our bodies has translated into the evolution of behaviors that make us unique (but not that unique).

In a conceptual sense, lessons learned throughout the history of evolutionary biology apply to human life and health and our understanding of who we are. But let's make it more personal. What have we learned from other animals that tells us about ourselves as humans? Specifically, what do other animals have to do with the exact genes that are inside our bodies right now, the ones that make us living, breathing, eating, sleeping, reproducing humans? It can't be that much, right? Maybe monkeys can tell us some things?

Monkeys ain't even the half of it.

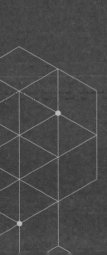

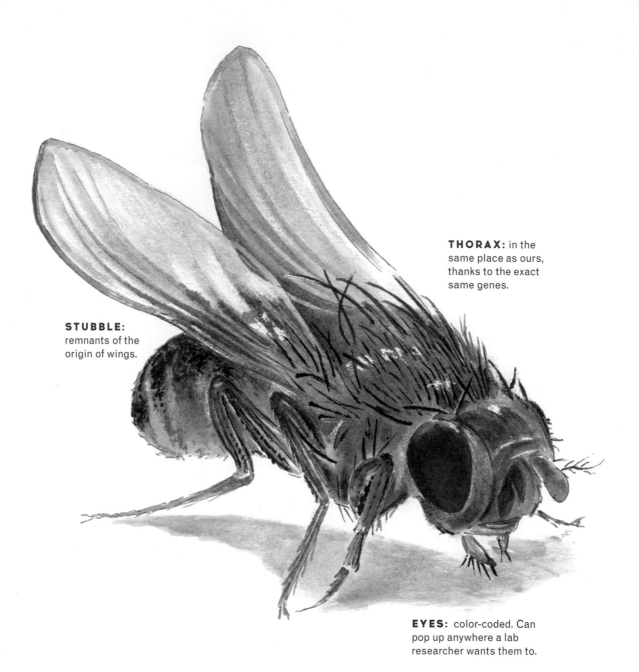

THORAX: in the same place as ours, thanks to the exact same genes.

STUBBLE: remnants of the origin of wings.

EYES: color-coded. Can pop up anywhere a lab researcher wants them to.

FAM-O-METER

44-61%

of our genes also show up in the *Drosophila melanogaster* fruit fly and vice versa. These genes are homologs, meaning they look the same because they probably came from the last common ancestor we two species shared. That ancestor would have lived a whopping 782.7 million years ago. We'll get into this more deeply, but—spoiler alert—that's a shockingly high retention rate.

75% of our disease-related genes have homologs in the fruit fly, which means even if we don't have a lot in common, we have enough in common to study *Drosophila* for potential genetic disease therapy.

SCALE
0.118 inches
(3 millimeters)

FRUIT FLY

(Drosophila melanogaster)

THE ORIGINAL SUPERMODEL
OF THE WORLD

What does the genome of a fruit fly have to do with me? Everything. The fruit fly *D. melanogaster* is one of the most studied model species in the history of science. As such, it has provided some of the most important breakthroughs in our understanding of genetics. And one of the headiest concepts we've come to understand? How genetically much we have in common with a fruit fly.

WHEN IN (PRE)HISTORY/
WHERE ON THE RIVER

Drosophila is one of the most studied animals in the history of science, but science still comes up empty on the little guy's prehistory. We do know that in the long time it's been around, it has certainly gotten around: it's closely related to an oddly high number of other insects on the planet and lives on every continent except Antarctica. There were two distinct *Drosophila* grandparent populations that come from Europe and Africa. But so far, that's about all we know.

Normally, fossilized bones might offer some clues to an animal's heritage. But since it has no bones, an insect only becomes a fossil if it falls into ancient amber or leaves a print in clay, and even then the fossil has to last a billion years and turn up someplace where a human will find it.

Humans simply haven't found any fossils that speak directly to the origins of *Drosophila*. The oldest insect fossil we have is about 400 million years old. It looks a little like a mayfly or dragonfly, and it could be related to most insects on Earth today.

But the emerging study of molecular clocks is offering some leads. A group of researchers in England led by Koichiro Tamura have developed an algorithmic way of tracing patterns in DNA that might be associated with the accumulation of new chunks of DNA over time. They've even turned their system into software called MEGA. According to MEGA, the *Drosophila* we know and love emerged between 26 million and…192 million years ago. A pretty mega margin, so we'll keep honing and hoping.

Darwin collected so many specimens that scholars found them squirreled away in various museums and libraries for decades after he was gone. Such was the case for his pair of *Drosophila*, not attributed to his collection until 1936. He couldn't have known the species would become living proof of many of his theories, especially two that he struggled with and never solved himself. In fact, the researcher who made the fruit fly famous set out to prove Darwin wrong about everything.

BEASTLY BREAKDOWN

EYES

In the early 1900s in Massachusetts, researcher Thomas Hunt Morgan, a graduate student at Harvard, was convinced that Darwin was wrong about inheritance being linked to male or femaleness. He thought it was foolish to think Mendel's work with dominant and recessive traits could apply to animals. Mendel worked with peas. Animals aren't peas.

Morgan had worked with fruit flies in college. They're small, and a single fertile mating pair can produce hundreds of babies in 10 to 12 days. (That's better than mice, which have a handful of babies every 3-4 months.) He noticed that some of them had white eyes and some had red—like Mendel's wrinkly and smooth peas, but animal. Morgan decides that *Drosophila* is his key to proving how traits really get passed down. He begins taking incredibly detailed notes about which flies are mating with which, keeping them in special cases and marking them with teeny-tiny dabs of fly paint.

After a few generations, Morgan's flies start to show a pattern: The females get white eyes only if both of their parents have white eyes. Males can have white eyes if only one of their parents had white eyes. *Dammit, maybe Darwin was onto something.* Over the next several years he conducted other experiments, just to be sure. In 1915 Morgan published the landmark work *The Mechanism of Mendelian Heredity*, the first hard evidence "that chromosomes act as the carriers of inheritance."

Morgan kept proving this for the rest of his career. He established the now famous Fly Room at Columbia University: all *Drosophila* all the time. In 1933 he won the Nobel Prize in Physiology or Medicine "for his discoveries concerning the role played by the chromosome in heredity." The speaker who delivered the prize explained Morgan's work to the assembly by giving shout-outs to Darwin, the pea-splicing work of Mendel, and a long passage devoted to a lesser-known contributor that started: "Another cause for Morgan's success is no doubt to be found in the ingenious choice of object for his experiments. From the beginning Morgan chose the so-called banana-fly, *Drosophila melanogaster*, which has proved superior to all other genetic objects known so far."

Fast-forward to 1993. A geneticist in Switzerland working on mice isolated what he thought was the "master control" gene for eye development, which he called Pax6, because it codes for the Pax-6 protein. He injected the

Pax-6 protein into a *D. melanogaster* embryo. And lo, at every injection site, the fly grew a tiny, red, compound eye. Not a mouse eye, a fly eye, which means Pax6 was indeed the "make an eye" gene. And not just for mice, but for flies and humans and everyone else who'd ever seen for 600 million years. Long live Pax6.

THORAX

The thorax is the middle section of an insect's body, in between the head and the abdomen. If the insect has wings, it's where the wings emerge. In 1915 in Morgan's Fly Room, a

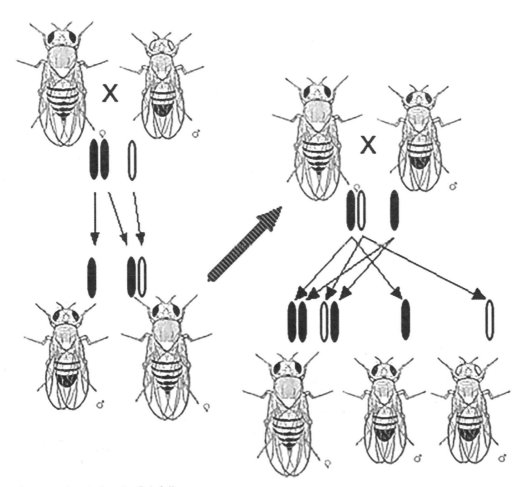

Over a century before the fly's full genome was mapped, this 1919 diagram outlines the pattern by which fruit flies exchange genetic information. From Thomas Hunt Morgan's work *The Physical Basis of Heredity.*

student found a spontaneously emerged mutant that had two thoraces instead of one, for a total of four body sections instead of three. Everyone was very excited. They named the fly bithorax and started breeding him, hoping to isolate the cause of the mutation.

Over the next century, researchers found that they could create mashups of all of *D. melanogaster*'s body parts: cause legs to grow where eyes should be, two pairs of wings, two tails, or no head. It was notable, they realized, that they were able to make body parts appear, but each part was always complete. Whatever instructions they were playing with, the body held the information in chunks.

Playing Frankenstein led Christiane Nüsslein-Volhard and her team of researchers to make one of the most important discoveries in this history of biology. They isolated the group of genes responsible for what goes where, a string of 180 base pairs responsible for building sections of *D. melanogaster*'s body: one head, one thorax, one abdomen. Within the 180, they discovered that a smaller set of three controls *D. melanogaster*'s body order in the embryonic stage. That is, before the animal is born, in its little fruit fly egg, those three genes are working away, turning stem cells into a fly, and in a specific order: head, thorax, abdomen. They called these genes HOX genes. The biggest deal of all is that those same HOX genes show up in most animals and animal embryos and control the same functions in the same order (all bilateral animals at least, as compared to radially symmetrical animals, like starfish, jellyfish, or sea urchins).

From *Contributions to the genetics of* **Drosophila melanogaster, by Thomas Hunt Morgan et. al., 1919.**

This discovery was a revelation at the time. Creatures that are symmetrical down the middle with a head and a rear, from horseshoe crabs to humans, have HOX genes, even though they vary in number. It was the answer to Morgan's question about traits being linked. This discovery was the beginning of our modern understanding of developmental biology, and it connected developmental biology to its origins, back through millions of years and across millions of species. Discovering HOX was like discovering a living Rosetta Stone, a single genetic template with which to start translating evolutionary histories for animals across the kingdom. The breakthrough was so massive that it opened up an entirely new field of study: evolutionary development, or

evo devo for short. In 1995, Nüsslein-Volhard and her team won a Nobel Prize for "their discoveries concerning the genetic control of early embryonic development."

"STUBBLE"

In 1901 Morgan was cataloging as many traits of *D. melanogaster* as he could. He noticed that some of its traits seemed to be linked: notched wings and thick butt bristles, for instance. A fly almost never has one trait without the other. The finding inspires one of his students to stay up all night mapping out the connections in a rudimentary (but accurate) map of the fly's chromosome. It's the first real attempt at mapping a genome, drawn by hand.

In 1988 a group of Canadian researchers find that the same gene responsible for butt

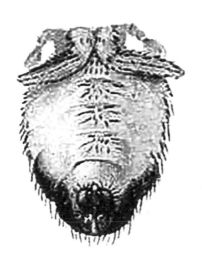

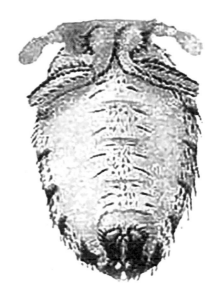

bristles is also tied to the development of wings and legs. They name the gene stubble.

The genetic connection between stubble and wings supports a new theory about the evolution of flying insects. They may have originated on the water. These prehistoric proto-insects had a bunch of simple appendages, some in front, some in back, some in between, and all good for swimming. But slowly, some insects traded swimming for skimming around on the water's surface. In the air, some appendages became legs, but others were more helpful catching a little wind. Skimming turned to gliding. Gliding to flying. And as wings grew, butt stubble became vestigial.

MYOSIN

Whenever any animal moves, it's because of the same teeny-tiny molecule (technically a long string of proteins, which is called a polymer) called myosin. Small insect wings, including *Drosophila*'s, which beat several hundred times per second, have the fastest myosins in the world. Cheetah's myosins are up there, too, but generally smaller is faster, and bigger is slower.

HEART

In 1993 researchers noticed that certain fly embryos in their lab just never developed a heart. They compared the DNA of the healthy flies to the heartless ones, and were able to isolate the heart-making gene. They name it tinman, because, well, who doesn't love *The Wizard of Oz*? Soon hundreds of scientists are studying tinman. The more researchers study it, the more certain they become that the *Drosophila* heart is basically, fundamentally, chemically very much like the human heart. Not so with other flies, even other fruit flies.

For example, the chemicals that make *Drosophila* hearts speed up also make human hearts speed up. Same thing for slowing the heart down. Today, *Drosophila* is the first test subject for certain heart medications that end up in our human bodies.

BRAIN

Some say the brain is more complex than the heart, but human/fly analogies exist there, too. If you want to find a human gene causing neurons to develop or degenerate, use the fruit fly as a road map. Studying fruit flies has helped humans better understand Alzheimer's and Parkinson's. In 2017, the researchers who isolated the genes that control circadian rhythms in *Drosophila* won the fly its sixth Nobel in physiology/medicine. Their findings showed how regular intervals of sleep and wakefulness are present in much of the animal kingdom, and that those rhythms have deep genetic roots and disrupting them can have significant health implications, from organ stress to weight gain. On the flip side, mutations in circadian genes can cause individual animals to have rhythms that make no sense within their populations or habitats. If you stay up late every night but can still get up early, you might be a circadian mutant, too.

ALCOHOL TOLERANCE

Speaking of hearts and brains and staying up too late: fruit flies like to drink. Or rather, they like the effects of fermented fruit juice. Drunk *Drosophila* flies fly crooked and fall out of the sky, and some become more aggressive—females will even head-butt each other. Males turned down for mating by female flies are more likely to seek out alcohol (though they may be attempting to get rid of unattractive parasites, rather than sad fly feelings). By the 1990s, researchers had isolated the gene that caused certain flies to get drunker faster: they call this version of the gene cheapdate.

Cheapdate flies also often show a gene pattern named amnesiac, so named because it seems to contribute to a drunk fly's forgetfulness about things like the location of food and consequences of drinking. And yes, we humans have these same genes in us, which may contribute to alcohol addictive behaviors. If researchers can interrupt those genes from working in flies, perhaps there's hope they can also help humanity handle our liquor better.

Am not I
A fly like thee?
Or art not thou
A man like me?

For I dance
And drink and sing,
Till some blind hand
Shall brush my wing.

— excerpt from
William Blake's "The Fly"

LESSON OF THE FLY: Very personal traits aren't personal at all. We share many traits with most of all animals on Earth.

TERMS DEFINED: HOX genes, Pax

RETINAL STALK: no retina nor eye, even

VENT SHRIMP

(Rimicaris exoculata)

EYELESS WONDER

In thinking about all the genes that trigger the formation of traits, it's worth noting that those same genes can trigger the lack of formation of a trait.

Just discovered in 2012, tiny deep-sea vent shrimp live in the near-boiling water around volcanic thermal vents in the sea floor. And they have evolved to have no eyes. Losing one's eyesight or eyes is not unusual for animals that live in the dark. There are even other very similar species of shrimp with very diminished eyes.

But these shrimp are unusual in that they do have retinal stalks, the bundles of nerves that usually connect eyes to the brain, but no eyes. Naked retina, evolutionary ophthalmologists call them. But the retina are in fact active and in a way provide more protection than the name would suggest. They sense light changes via a photo-sensitive patch of tissue on the shrimp's back. Thus, the vent shrimp can adjust its movements according to what's going on with the vent, protecting itself from the extreme heat and sulfuric sputterings without exposing a single naked eye.

LESSON OF THE VENT SHRIMP: Once again, the genome comes first; the traits come second and sometimes not at all.

GREAT BARRIER REEF DEMOSPONGE

(Amphimedon queenslandica)

YOUR SIMPLE COUSIN

INHALANT PORE: used for sucking in food and reproductive tissue

EXHALANT PORE: used for expelling waste and offspring

"MUSCLE": (aka muscle cells, aka collagen, the same stuff our muscles are made out of, just arranged quite differently)

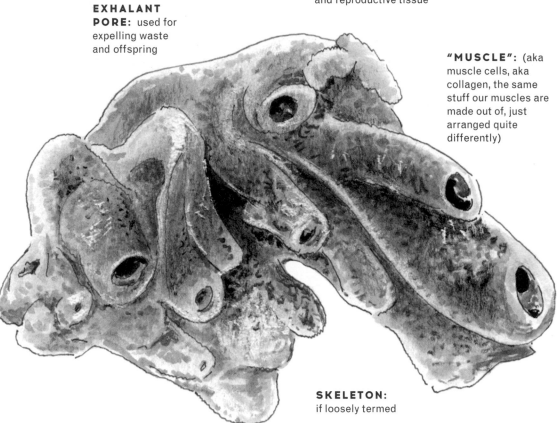

SKELETON: if loosely termed

SCALE
It all depends. Usually no taller (or rather deeper than) 1 foot (30 centimeters), but they can grow to cover areas of several feet.

FAM-O-METER

60%

Demosponges are pretty simple. And to put it way too simply, their bodies consist of far fewer tissues, structures, and general genetic goings-on than the human body. As such, the amount of our DNA that has direct parallels in the demosponge accounts for around 60% of our genome. But get this: A whopping 70% of the *demosponge*'s genome has parallels in our genome. They're us like we used to be.

This animal barely looks like an animal, and it even leaves something to be desired as a sponge. But thanks to a common ancestor, the great barrier reef demosponge is probably our simplest living relative. We share our most fundamental genetic building blocks with it. Our shared genes are so fundamental, in fact, that there's really not a whole lot more to being an animal than what the sponge has going. It's got the basics covered. All the bells and whistles we other animals have, that's all just kinda fluff.

ALL IN THE FAMILY

The notable stat here is the other way around: 70% of a sponge's genome matches ours. What? You might/should be saying: *How can this seemingly shapeless marine invertebrate have more in common with us than we have in common with an opossum or a platypus, which are actual cute animals with hair and legs and eyes and babies?* Genetic overlap happens in ways that we can't see. We just fixate on the trivial stuff: eyes, legs, a nice laugh.

Does this mean that humans evolved from sponges? No, it does not. It can't, because sponges are still around, and so are we. Again, that's like asking if your cousin is your great-great-great-grandfather. Rather, it means that we share a common ancestor, just as we do with all animals in this book. Except that with sponges, that ancestor was very, very early in animal evolution. And the sponge hasn't added as much since then.

Every time we sequence a genome, we're adding pieces to the puzzle of life on Earth. It's a puzzle with an indefinite number of pieces and absolutely no picture on the box. Our only frame of reference is to compare the genomes of different organisms to one another: see what they have in common and what they don't.

There are two major ways that genomes can overlap:

Analagous genes (genetic analogs): chunks of DNA that *did not* come from the same place but now serve a similar purpose in the modern animal. (For example, making nerve connections has evolved in a few different ways through time.)

Homologous genes (genetic homologs): chunks of DNA that came from the same place (share a common ancestor) but may *or may not* serve the same purpose in the modern animal. (For example, unlikely telltale patterns of DNA may show up in related animals, but...it's hard to tell what their job is.)

As you read onward, keep your eyes peeled for spin-offs: homologs that come about via reproduction vs. homologs that come about because of changes in the genome itself (mutation, gene duplication, horizontal gene transfer)—but that's a story for another day.

For now, think only of this. Long ago... when the world was young...there was a very simple animal, great great great (etc.) grandmama to you and the Great Barrier Reef demosponge. That ancestor was very, very early in animal evolution. And the sponge hasn't added much since then.

WHEN IN (PRE)HISTORY/
WHERE ON THE RIVER

For this animal, we're going way back. To the beginning of multicelled animals. The sea sponge is related to the 1.5 million known species of animals that we think of as animals, the metazoans. If all mammals have hair and nurse their young, all metazoans check all (or most) of the following boxes:

- "Eat" other living things in order to live
- Breathe oxygen
- Move

- Reproduce with the help of a partner (okay, but yes, we see you, whiptail lizard)
- Grow from a clump of cells, the blastula, during embryonic development

Metazoans are all of us. Parrots. Shrews. Anything more complex than a protozoan.

Today, there are over a hundred species of demosponges worldwide, living in every body of water, salty and fresh, tropical, arctic, and at the bottom. The ancestors of the modern-day sea sponge emerged like so:

2500	541	MILLION YEARS AGO	65
PROTEROZOIC	PALAEOZOIC	MESOZOIC	CENOZOIC

o

 ...950 million years ago, they split from the lineage they share with fungi.

GREAT BARRIER REEF DEMOSPONGE

 ...750 million years ago, they split from the lineage that would become comb jellies.

...800 million years ago, the proto-sponges' lineage split with the lineage that would become cnidaria (the group containing most of what we think of as jellyfish—moon jellies, box jellies, purple jellies—as well as corals, hydras, and Portuguese men-of-war), reptiles, amphibians, birds, and mammals.

BEASTLY BREAKDOWN

REPRODUCTION

Demosponges usually reproduce hermaphroditically. One sponge expels sperm from its exhalant pore, and the sperm swim around a bit until they wash into another sponge's inhalant pore. Some sponges also expel their young as eggs, and some incubate the embryos inside their bodies. Here again, the demosponge seems to represent a blank slate of multicellular organisms, keeping its options open.

DIGESTION

Sponges eat and breathe much the same way they get pregnant: they use little moving fibers (flagella) to wave plankton and oxygen rich water into their internal chambers, digest their food right inside their very cells, and expel waste through their—you guessed it—exhalant pore.

BODY PLAN

If you can believe that we share a body plan with a fruit fly, perhaps it's less of a stretch by now for you to believe that we share some body-plan-making DNA with sponges. No, their body plans do not look like ours. Remember, they split from us before we split from jellyfish and anemones, whose bodies are arranged on a radius (radial symmetry) rather than a bias (bilateral symmetry), like us and fruit flies. But the fact of having a body plan at all is something that all metazoans have in common. And some of what we share with sponges are those signaling genes.

TISSUE

Tissue is made up of many cells of a particular kind: muscle cells make up muscle tissue; bone cells make up bone tissue. Those specialized cells come about and arrange themselves correctly because DNA tells them to. It tells other organic molecules how to work together to make and become proteins, which come together to make cells and then tissues. Every tissue in your body, from brain to skin, came about via a genetic set of instructions, and those instructions all exist in the demosponge, too, instructions for cell development, cell growth, cell specification, cell death. Somatic cells, and germ cells, and the way cells stick to each other—all of this is in the demosponge, too. They start, as we do, as a blastula—a familiar word to anyone who has ever had to undergo fertility treatment. We and sponges both start as a fertilized egg that begins dividing and multiplying, and all of this information and more is already contained within the genome.

SKELETON

That's right. These blobs of what looks like mostly negative space have skeletal planning genes, parallel to the ones in humans and fruit flies alike. Only instead of a bunch of bones and cartilage, they have just two components to their skeleton. One is something called a siliceous spicule, which sounds like a villain from a Harry Potter story but is basically a little spike that holds the rest of the blob upright.

The second is called spongin: a special type of collagen that either squishes outward and spreads the sponge's body mass, or forms into large fibers, like scaffolding for the goo. Both of these are made out of chitin, the same stuff that makes up the exoskeletons of arthropods like bees, fruit flies, spiders, giant marine isopods, lobsters, and crabs.

COLLAGEN

Sponges don't have muscles, although they have genes that, in humans and other animals, look like genes for coding muscles. That is, their genes send up a directive that in humans looks like "make a muscle!" but instead their bodies make collagen that links their variety of cells together and allows them to have become one multicelluar organism diversified and complexified over time. It simply took on new functions in animals like us. Throughout biology, you'll hear the phrase, "[thing] is shared by multicellular organisms." Rather, these genes are the reason for multicellular organisms.

SYMBIOSIS

No organism is an island. Coevolution is a reality for every organism on Earth, but sponges are especially good at living in harmony with other organisms. This is a biological process called symbiosis, wherein both animals in the relationship benefit from their teamwork. One Mexican species of demosponge proved to be home to more than 100 other species of animal and algae (small aquatic plants). And this doesn't even include bacteria.

The details of the relationship between sponge and whoever's sponging off of the sponge will vary from symbiosis to symbiosis, but it seems that the demosponge is an especially gracious host: it emits various organic chemicals that are useful to other organisms, like growth hormones.

Science is just starting to unpack symbiotic exchanges like this. It's relatively recent news, for instance, that the billions of bacteria living inside the human body are really *part* of our bodies and have actually contributed to our evolution into the complex beings we are. The line between human and human-borne bacteria is not a hard line, and the two should not and cannot be separated. The sponge and its extreme symbiosing offer a window into how this process might have looked early on, when animals were simpler and the transaction was more like, "Let's live near each other and swap useful juices." But even in this simpler form, demosponges are—genetically, chemically—speaking the same language as we are. Pharmaceutical researchers have taken to farming demosponges as an easy source of organic chemicals, just like the ones that become active ingredients in some human medications.

IMMUNITY

One reason that symbiosis became possible for the sponge (and all other animals) is the very basic ability for an organism to tell its own cells from cells of someone or something else. This ability is called allorecognition, and it's the first step to building a complex immune system. It shows up in the sponge's genome. The sponge doesn't really have much more of an immune system than that, but, as its knack for symbiosis shows, it seems to be working out well for it. Similar to its lack of bones, muscles, and neurons, the demosponge has the groundwork, which it inherited from a long-ago ancestor it shares with us. It just hasn't built the house. Or rather, it built a different house, open-air, where everyone's invited.

PROTO-TUMOR SUPPRESSORS

Many of the same genes that helped bring about multicellular organisms are implicated in cancer. Cancer in all its forms is the result of too many cells continuing to grow in an area where they shouldn't. The demosponge possesses gene markers reminiscent of tumor suppressors in more complex animals—genes with messages to stop growing. But the sponge is almost nothing but growing cells, so cancer is, as far as we know, a nonissue.

NEURON GENES, NO NEURONS

Sponges have no neurons or nerves at all, not even really any discernible internal signaling system. They either never had them or they lost them somewhere early in the history of life. But they do have genes that, in humans, help code for neurons.

What? Why? What do sponges do with neuron codes if they don't have neurons? Those codes indicate that sponges have some chemical self-signaling systems going on, which of course they do, as animals. Or at the very least they have the inert potential for it. Perhaps more interestingly, if the common ancestor of sponges and humans had genes that, in humans, make nerves, what did the common ancestor use them for? And what does the sponge do instead? How does it understand its environment? These are questions still begging for research.

CIRCADIAN RHYTHM

Genomic studies suggest that circadian rhythms have evolved a few different times in evolutionary history. And you might have imagined that our genetic circadian rhythm originated in recent relatives, yawning in the treetops as the sun goes down. Or at least an animal relative that has eyes to close or a body to lie down and rest. But similarities in the sponge genome and human genome suggest that the set of circadian genes running through our human lineage may have originated in our common ancestor with the sponge. It may have simply been a product of the proto-sponge's squishy, formless body being exposed to regular patterns of sunlight exposure over the course of each day.

LESSON OF THE DEMOSPONGE: The basic building blocks of our animal lineage are even more basic than you could have imagined.

TERMS DEFINED: Symbiosis

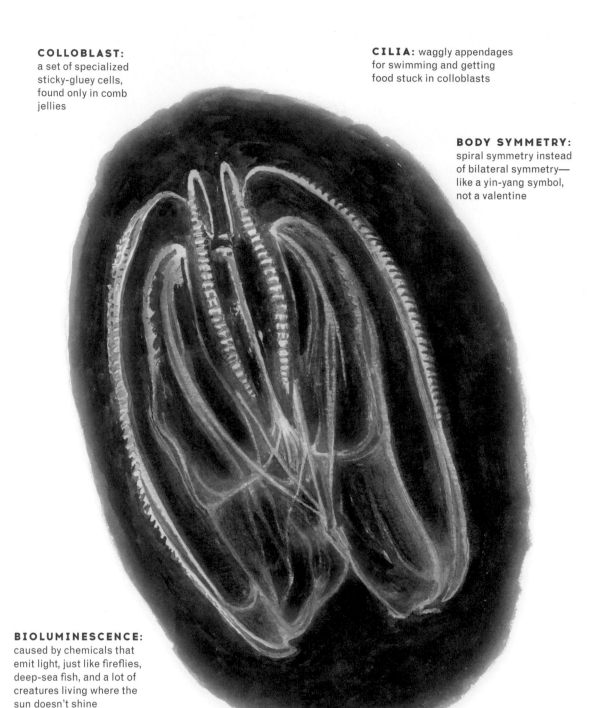

COLLOBLAST:
a set of specialized
sticky-gluey cells,
found only in comb
jellies

CILIA: waggly appendages
for swimming and getting
food stuck in colloblasts

BODY SYMMETRY:
spiral symmetry instead
of bilateral symmetry—
like a yin-yang symbol,
not a valentine

BIOLUMINESCENCE:
caused by chemicals that
emit light, just like fireflies,
deep-sea fish, and a lot of
creatures living where the
sun doesn't shine

SCALE
3–5 inches (7–12
centimeters)

FAM-O-METER

?

For reasons
explained below:
it's complicated.
Please also refer
to the sponge on
page 147.

COMB JELLY
(Mnemiopsis leidyi)

YIN AND YANG—SOME OF THIS, NONE OF THAT

The debate between demosponge researchers and comb jelly researchers raged for many years, with both arguing that their research animal was the most recent common ancestor with humans. The debate has since been settled (it was the demosponge), but the comb jelly still presents an interesting genomic puzzle. The comb jelly's genetic traits are illuminating, but the traits that it *doesn't* have, especially when compared to the demosponge, further flesh out the genetic story of life on Earth.

WHEN IN (PRE)HISTORY/ WHERE ON THE RIVER

Comb jellies swim all over the world, including Antarctica.

They're often mistaken for jellyfish, but jellyfish are radially symmetrical, like a pinwheel, and are in a totally different part of the river of life. Comb jellies are rotationally symmetrical, like a yin-yang symbol. According to one American sponge researcher, categorizing a jellyfish and a comb jelly together is like putting a platypus "in the duck box or a beaver box." Whereas jellyfish pulse through the water mouth downward, their umbrella-like bells opening and closing, ctenophores like the comb jelly swim mouth forward; in most textbooks and scientific papers, they are depicted the wrong way up.

BEASTLY BREAKDOWN

NEURONS BUT NO NEURON GENES?

Comb jellies have nervous systems, though their neurons aren't as specialized as those in the human body.

What they do have are the genes other animals use to build neurons and neurotransmitters. These genes are called micro-RNA, so called because it's their job to build and fix RNA. RNA's job is to tell the rest of the animal's body how to build and fix itself.

But if comb jellies branched off before sponges, it could mean that they evolved nervous systems independently from all other animals, including us. This laid to rest the question of whether sponges or comb jellies deserve the distinction of our earliest relative. If sponges are the earlier of the two clades, the story unfolds neatly: they had the genetic building blocks for a nervous system, which ctenophores elaborated and bilaterians went to town on. This narrative shatters if ctenophores branched off first.

Here's what comb jellies *do* have.

COLLOBLASTS

Predatory Glue Traps: Ctenophores collar their prey with glue-secreting cells called colloblasts, which are unique to them.

CILIA

We use cilia to clear mucus from our airways. Ctenophores use theirs as flippers, sense organs, even as teeth. One species of comb jelly, the sea walnut, uses its cilia to create imperceptibly subtle water currents that draw fish and other

Neurons: When we think neurons, we often think "brain cell," but a neuron is any type of signaling tissue. The human body has four types: brain neurons (of which there are still more specific types), sensory (which feel stuff), motor (which move stuff), and interneurons (which help the others connect).

prey into its mouth. It is such an effective predator that whenever it enters a new body of water, it throws the food webs into disarray. In the 1980s, it got into the Black Sea, where it devastated anchovy populations, to the detriment of the local dolphins. For something with the destructive power of a mini Charybdis, it sure is beautiful: the cilia are so delicate and clear that they catch the light like little prisms, creating a rippling rainbow effect.

LESSON OF THE COMB JELLY (AND DEMOSPONGE): Here we
have two animals that seem equally different from humans. Both are simple. Both live decidedly unhuman lifestyles. But by looking at exactly which of the simple traits each animal possesses, we get a snapshot of the simple forms from which we arose—and the simple forms from which we did not.

THROUGH DARWIN'S EYES

The mouth is situated in centre of square funnel shaped projection, which becoming narrower forms the gullet.—[note (b)] Sept: 2d The situation of mouth is strongly marked by a black dot: it always appears closed.—[note ends] The situation of the mouth, as before mentioned, is in rather a deep depression;—the edges of this contract very suddenly if touched; & I suppose by this manner any minute object is caught, which may afford support to the animal.—

Above, Darwin is describing a comb jelly, but in another publication, he describes an invertebrate called the tunicate next to a comb jelly, showing that he's decided the two are closely related. He later comes clean about this and adds, "They were not in fact tunicates, but were comb jellies, ctenophores of order Cydippida." He wasn't far off, however. Another marine invertebrate, the tunicate, tells a similar story. Just as the comb jelly has nerve genes but no nerves, the tunicate has no spine or brain, but does have the genetic code for what in humans becomes a spinal cord.

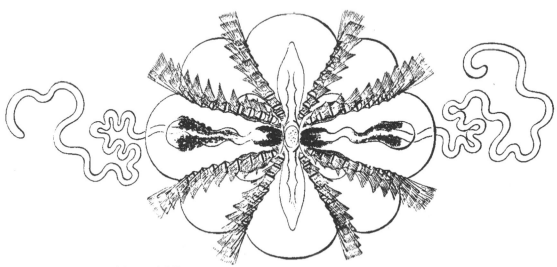

Top-down view of the comb jelly *Haeckelia rubra*, by its official discoverer and namesake, pioneering artist and marine biologist Ernst Heinrich Philipp August Haeckel.

SOUTH AMERICAN LUNGFISH

(Lepidosiren paradoxa)

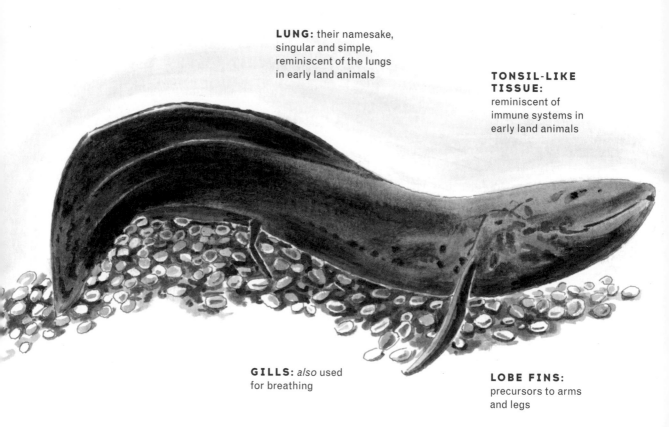

LUNG: their namesake, singular and simple, reminiscent of the lungs in early land animals

TONSIL-LIKE TISSUE: reminiscent of immune systems in early land animals

GILLS: *also* used for breathing

LOBE FINS: precursors to arms and legs

SCALE
4 feet (1.25 meters)

FAM-O-METER

63%, give or take a few hundred thousand transposons.

AFRICAN COELACANTH

(Latimeria chalumnae)

FOSSIL OR FAM?

Science tells us that every animal comes from the sea. But beyond that, we get picky. For many years, another heated controversy among evolutionary biologists was around whether the coelacanth or the lungfish was the closest living relative to humans' long-lost fishy predecessors.

LOBE FINS: those precursors to arms and legs again

SOUTH AMERICAN LUNGFISH

(Lepidosiren paradoxa)

CLOSEST (FISH) COUSIN

Lungfish are a group of fish characterized by the fact that they, as you may have guessed, have lungs and lobed fins. But be wary of the word *fish*. (And of the word *lung* for that matter.)

WHEN IN (PRE)HISTORY/ WHERE ON THE RIVER

Lungfish were abundant and widespread in the Devonian period, around 400 million years ago. Today, there are only six species left.

The Queensland lungfish split from the rest of the lineage about 270 million years ago. Certain aspects of its genome more closely resemble the coelacanth than other lungfish, especially in genes related to the brain.

Like the coelacanth, the lungfish appears in the fossil record as far back as 80 million years ago. Back then, they were fish in the sense that they swam underwater. And were not yet a fish, as evolutionarily different from fish of today as they are from us. More so, in fact.

First with fossil evidence and later via genome analysis, we now know that an early version of the modern lungfish shared an ancestor with modern land animals. An early lungfish was not the common ancestor itself, but shared a last common ancestor.

That means, as far as fish go, the lungfish is our closest living relative. And it can tell us a lot about ourselves.

AUSTRALIAN LUNGFISH

It was about 100 million years ago that the South American lungfish diverged from the Australian lungfish, which displays the most primitive characteristics. It has more prominent fins and will sometimes swim up to the surface to gulp air into its simplified lungs, should its gills not suffice. It pays to have options.

BEASTLY BREAKDOWN

LUNG

As its name suggests, the lungfish has a single lung—a long sac along the length of the body cavity—which supplements its usual gas exchange through its gills.

GILLS

Modern lungfish have external gills. Ancient lungfish probably started out with gills and developed the proto-lung over time. The increased access to oxygen may have contributed to the changes that were to follow.

LOBE FINS

The contest between lungfish and coelacanths to see which is more related to humans all started because of the lobe fin. *Lobe fin* describes forefins like that of the coelacanth: hefty, muscular, rounded fins. It was a fish with fins like these that first made the transition to land, pulling itself out of the water on scrappy little forelimb stumps that eventually became the first legs.

Because of those fins, the coelacanth was an early front-runner in a paternity dispute worthy of a daytime talk show. Besides, the coelacanth seemed as though it hadn't changed at all in millennia, and animals that appear not to have changed are much more likely (we thought) to represent transitional species. In short, because it looked old, the coelacanth was voted Fish Most Likely to Be Your Grandma, basically.

But in 2013, an international group of researchers found that a genome comparison among the West African lungfish, the African coelacanth, the chicken, and a handful of mammals proved once and for all that the lungfish had more in common with the rest of us than did the coelacanth.

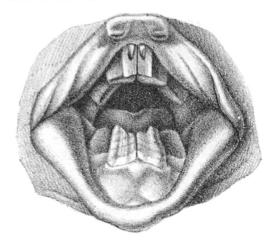

This 1856 look into the mouth of a lungfish highlights its "fused teeth plates," solid chunks of bone whose evolution predate socketed teeth, by French naturalist François-Louis Comte de Castelnau.

"Through a phylogenomic analysis, we conclude that the lungfish, and not the coelacanth, is the closest living relative of tetrapods (land animals)," reads their article.

(You might not think of a scientific paper as sounding sassy, but you know it when you see it.) More support for this conclusion came in 2017, when Japanese researchers contrasted the lungfish genome to both living cartilaginous fish (rays and sharks) and modern ray-finned fish, finding that the relationship between ray-finned fish and lungfish told the same story.

IMMUNITY

One of the adaptations that helped vertebrates thrive is our efficient immune systems. This is the system by which our bodies recognize foreign invaders and react to protect us. It does this mostly via the lymphatic system. Lymph nodes (those weird squishy bundles of tissue in our armpits and groins and throughout our body, the ones that swell up when we get sick) communicate and connect to one another via a bodily fluid called lymph, which is one of the clear liquids that make up our blood. Lymph carries bacteria and viruses to the lymph nodes, where it destroys them and makes antibodies to remember to destroy that bacteria or virus the next time it finds it. We also have secondary lymphoid tissues, like tonsils, adenoids, and Peyer's patches, which line the small intestine. A 2015 international study found those secondary lymphoid tissues in the African lungfish. This is more evidence that lungfish are close relatives to that fishy predecessor of all tetrapods. Not only that, it means that lymph-based immune systems pre-dated the emergence of tetrapods from the prehistoric seas. If you're gonna crawl out of the water, you'd better be ready for all the new germs out there.

The Jawbone Saga, or, There's No Such Thing as a Fossilized Ear: Before there was genomics, we studied a *lot* of jawbones. The details of jaws that were connected to the top of the skull versus jaws that weren't and what the connection of the jaw did for the ear bone or the brain case vary from animal to animal. These days, information that the jaw used to give us has largely been displaced by genomic information, which is to say that genomic information has mostly supported previous conclusions drawn by jawbone and ear-bone experts. Where this cache of jawbone knowledge *does* still come in handy is in fossil identification. We can't sequence genomes on most fossils: the DNA is too degraded or contaminated to be readable. But we can always go back to the jawbone archives and make comparisons there.

LESSON OF THE LUNGFISH:

Some of our earliest ancestors were fish-like fully aquatic animals. Deal with it.

AFRICAN COELACANTH

(*Latimeria chalumnae*)

OLD ENOUGH TO KNOW BETTER

In 1938, a South African fisherman pulled a crawly looking, blueish, incredibly muscular fish from the ocean near Sulawesi, Indonesia. He had seen a lot of fish, but nothing like this, so he brought it to his local university.

A professor there had seen fish like it before— but only preserved in a fossil from 80 million years ago. Yet very, very old does not mean the same thing as living fossil. There's no such thing as a living fossil. There is such a thing as a very, very ancient genome that doesn't appear to have changed a great deal in the past many million years, though. And that's the coelacanth.

WHEN IN (PRE)HISTORY/ WHERE ON THE RIVER

As with the sponge/comb rivalry, it's complicated.

BEASTLY BREAKDOWN

LOBE FINS: Lobe fins are muscular fins with a single bone that connects to the rest of the body. Compared to ray fins, which modern fish have, this type of fin is a serious throwback, and visibly identical to the lobe fins on coelacanths' fossilized predecessors.

Lobe fins were the first fins. Some of them evolved into legs and feet on land animals, others evolved into other types of fins, eventually becoming modern fish fins. In coelacanths and lungfish though, they stayed the same. If it ain't broke, don't fix it. Or more like, if you don't have mutations that take you in another direction and you find that you survive just fine, you'll probably stay that way.

This idea of not changing confuses people when it comes to evolution. This is the basis for the term *living fossil*. Animals that are living fossils are animals whose lineages first broke from others millions of years ago, yet the extant animals still look the same. But the problem with this idea is that it's superficial.

Protein-coding genes, the ones that affect the visible changes in an animal, can evolve at a very slow rate. But even in the slowest-adapting animals, transposable elements appear to be active and show high diversity.

Unlike other genomic features, coelacanth protein-coding genes evolve significantly slower than those of tetrapods. But analyses of changes in genes and regulatory elements during the vertebrate adaptation to land highlight genes involved in several areas of the genome that are relevant to humans. Changes in the coelacanth's genes for immunity, nitrogen excretion, fins and tail, neural responses, and olfaction can illuminate facets about the story of early animal transitions from water to land.

LESSON OF THE COELACANTH: Slow and steady still persists just like every other living thing on Earth.

TERMS DEFINED: Living fossil

ZEBRAFISH

(Danio rerio)

FULL TRANSPARENCY

Zebrafish are another of those model species we keep talking about: small, prolific, and useful in the lab. That means we know a lot about them because we've studied them so much, and we keep studying them because we know a lot about them, which gives us new and better questions to ask, and so the cycle goes. Zebrafish have another great quality, too: they're see-through. This makes it really easy to compare what we see in their genome to what we see with our very eyes. And our eyes are a lot like a zebrafish's, as it turns out.

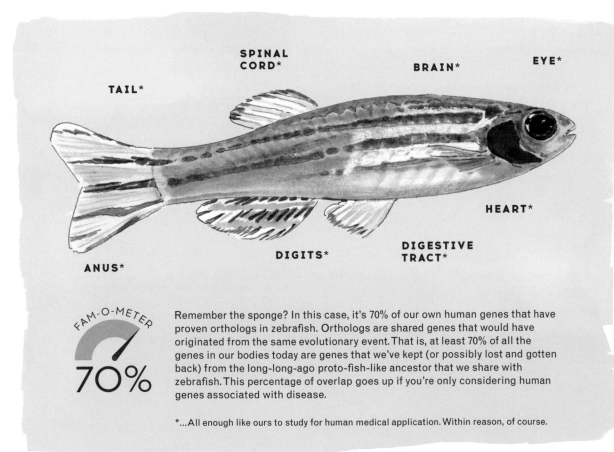

SPINAL CORD*

TAIL*

BRAIN*

EYE*

HEART*

DIGESTIVE TRACT*

DIGITS*

ANUS*

FAM-O-METER

70%

Remember the sponge? In this case, it's 70% of our own human genes that have proven orthologs in zebrafish. Orthologs are shared genes that would have originated from the same evolutionary event. That is, at least 70% of all the genes in our bodies today are genes that we've kept (or possibly lost and gotten back) from the long-long-ago proto-fish-like ancestor that we share with zebrafish. This percentage of overlap goes up if you're only considering human genes associated with disease.

*...All enough like ours to study for human medical application. Within reason, of course.

WHEN IN (PRE)HISTORY/
WHERE ON THE RIVER

Humans and teleosts split so long ago that we have little in common with most true fish these days. That means no one's testing medications out on fish before entering human trials, with the exception (to a certain extent), of zebrafish.

The demosponge of the vertebrate world, the zebrafish genome doesn't have a lot of bells and whistles. What they do have going on shares a lot of overlap with humans—sequences like those among homeobox genes: body plan, organogenesis, the way embryos grow from zygote (fertilized egg) to fetus to birth. Our organ systems are similar, and for the first five stages of embryological development, it's hard to tell a zebrafish embryo from a human one.

We know this because the zebrafish are platinum-level model species. They're small. They reproduce quickly. And not only are their eggs transparent—their *bodies* are, too. This makes it really easy to observe their internal organs under the influence of different experimental variables.

LESSON OF THE ZEBRAFISH:
See? Told you your ancestors were fish-like. Heck, you were fish-like from about week three to week six of your mother's pregnancy.

A zebrafish in its "larval" stage looks uncannily like a human embryo around four weeks after conception.

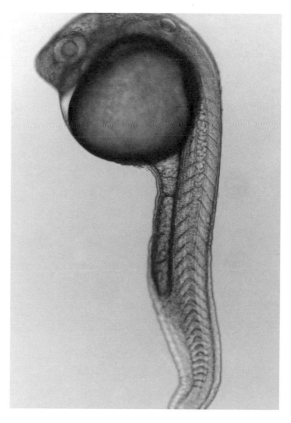

AFRICAN CLAWED FROG

(*Xenopus tropicalis*, *Xenopus laevis*)

SLEEPER CELLS

The first-ever amphibian to have its genome decoded was *Xenopus tropicalis*, which humans like because it's always been really good for testing human stuff. Its genome just confirms more of the same: genetically, *Xenopus* is everywhere, it's good at everything, and there's a lot more to it than meets the eye.

SEX ORGANS: pretty flexible, but pretty useful to the science of human health

CLAWS: an old-school trait for the frog

SCALE
1.1–2.2 inches
(28–55 millimeters)

FAM-O-METER

ROUGHLY

70%

of our DNA overlaps with that of *Xenopus*. The overlap covers genes that affect the heart, lungs, cell differentiation, embryology, toxicology, and more.

WHEN IN (PRE)HISTORY/ WHERE ON THE RIVER

The most important chapter in *Xenopus*'s evolutionary success story began in 1930. A British researcher exploring in South Africa discovered that African clawed frogs breed quickly and are disease-resistant and easy to work with. He also discovers that they begin laying eggs like crazy when injected with a pregnant woman's urine. It's a living pregnancy test—the million-dollar idea of every man's dreams. The researcher brought a collection of *Xenopus* to England and set up a breeding lab where he could continue his research; he wanted to make sure his test was reliable. By 1940, he was sure. As the Greatest Generation defeat the Axis powers and march home to begin making 2.3 babies for their three-bedroom, two-bath houses in the suburbs, African clawed frogs begin taking over the world. It's a parallel amphibian baby boom.

There were so many *Xenopus* jumping around that breeders began selling them for other reasons, too: as mail-order pets and specimens for anatomy classes. But by the 1960s, as new pregnancy testing technologies arose, the froggy bubble burst. Laboratories and pet owners alike intentionally released *Xenopus* into the wild, where it was free to propagate at will. They dispersed to new habitats and geographic areas, thrived in a wide variety of changing and hostile environments, could eat almost anything, and effectively thwarted a wide variety of predators.

And thus, *Xenopus* intrigued researchers all over again. *Xenopus* species are the most widely used research animal for vertebrate developmental, molecular, and cell biology—the mouse or lab rat of amphibians. By the way, there's so much variation in their gene pool that some species of *Xenopus* are as genetically different from one another as mice are from rats.

BEASTLY BREAKDOWN

LARYNX

For evolution researchers, the diversity among the 20+ species of *Xenopus* is like having 20+ windows into parallel evolutionary realities. All *Xenopus* genomes have enough in common to make for easy genetic comparisons. And they adapt quickly, so it's easy to find individuals that exhibit different traits. In 2013, researchers from New York compared the genomes of two divergent species of *Xenopus* that had noticeably different mating calls. They found that indeed the two calls had evolved via totally different pathways; their genomes coded different neural pathways connecting the larynx and the brain.

EYES

A 1999 study of the eye gene family PAX6 in *Xenopus* found that the frogs are particularly susceptible to mutations there. In rather horrific experiments, researchers duplicated a series of human eye malformations, including a condition called aniridia, where the patient's eye has no telescoping iris, just one big black, open pupil. Thank you for your sacrifice, *Xenopus*.

SEX DETERMINATION

Like the whiptail lizard and velvet spider, the clawed frog is polyploidy, meaning it might retain multiple chromosomes from both its parents. In the case of the African clawed frog, this happened more than once, ultimately amounting to an animal with four whole genomes repeated in its DNA. And all of those genomes, it seems, show up at some point in the life of a frog. So a lot of what we know about how RNA and DNA function is because of the *Xenopus* and its genome.

While science hasn't been able to directly prove that this system is the "why" behind *Xenopus*'s ability to succeed in a wide variety of environments, science does accept that the more genetic material an animal has to choose from, the more adaptable they might be—if, that is, the animal's genes are themselves used to dealing with information overload.

But *Xenopus* is able to do it. In at least one species of *Xenopus*, *X. laevis*, we know that it's polyploidy-ness happened many millions of years ago. This makes it ideal for the study of *poly*ploidy, but really complicates the idea of studying its one genome and applying that information to all other related groups, subspecies, and species. When one study compared *X. laevis* to *X. tropicalis*, for instance, findings showed that it did some fancy dancing as it evolved its genome. It maintained an intact version of one of its four genomes (subgenomes, actually), while the other genome did what genomes usually do from generation to generation: rearranged some base pairs, lost some base pairs, lost some whole genes. Some of the new changes showed up physically in the new frog; some didn't—basically, because it had two genomes to choose from. Best of both worlds. Changes without the risk. If the new genome doesn't cut it, you still have the old one to fall back on.

SEX ORGANS

Multiple sex-specific chromosomes mean multiple places for them to go wrong. In one famous study, an American researcher was hired by a chemical company to track the responsible disposal of industrial waste, only to find that it caused an entire population of clawed frogs with both male and female sex organs and extra limbs.

CLAWS

One doesn't associate claws with frogs, but *Xenopus*'s characteristic claws play a key part in their success. Any animals that don't have claws have evolved to not need them. For *Xenopus*, claws are an excellent defense against predators and prove useful in catching their favorite food: any bite-sized animal that crosses their path, living or dead. Their digits are also highly sensitive, which helps them locate food.

SENSES

While *Xenopus* does have an excellent sense of smell, it also uses an interesting way of detecting prey that works especially well in water. A lateral-line system is a connected line of nerve-packed holes along its nose that pick up both chemical and electrical changes in its environment. With this system, it can "taste" chemical changes and feel small changes in current or shifts in electrical energies and underwater. It evolved way back before the divergence of sharks from the rest of the animal lineage, and many modern amphibians and fish still use it.

Since 2013, a handful of studies have shown genetic correlation between smell, taste, and whiskery/feely/tactile systems in terrestrial reptiles and birds. They may even correlate to the inner-ear hairs in humans. These lateral-line gene markers appear in lungfish, remember. They may yet prove to relate to those mysterious non-neuron neural codes in demosponges.

THROUGH DARWIN'S EYES

Darwin doesn't seem to have cataloged *Xenopus*, whose range was considerably smaller just 150 years ago. When it came to puzzling over the relationship between evolution and geographic limitations, Darwin often looked to his friend and colleague A. R. Wallace, who wrote a two-volume book on the subject. Darwin had noticed the effects of limiting geographies on a small scale, of course, but global dispersal still confounded him. He and Wallace spent hours of discussion and pages of correspondence wrestling with the issue. Especially problematic were species that looked similar to one another but lived in very different parts of the world. Both Darwin and Wallace struggled their entire lives to understand how this could have come about.

In 1909, six years after Darwin's death, a zoologist named Hans Gadow contributed an article to a collection of essays revisiting the work of Darwin and Wallace.

In it, he complains about the inability of his colleagues in the biological community to differentiate species according to relatedness versus geography. He uses as an example two disparate but reminiscent species of frogs. "If these creatures lived all on the same continent, we should unhesitatingly look upon them as forming a well-defined, natural little group. On the other hand...," he complains, "Pipa in South America, *Xenopus* and *Hymenochirus* in Africa [look so much alike and are so obviously] one ancient group," that his colleagues readily "use their distribution unhesitatingly as a hint of a former connection between the two continents. We are indeed arguing in vicious circles." In 1912, just a year before Wallace's death, a young German researcher came up with a possible solution when he published about his vision of Pangea and the mobility of Earth's continents.

LESSON OF THE AFRICAN CLAWED FROG: When a polyploidy organism can access the DNA from all of its chromosomes, it can diversify so quickly that it blows the term *species* right out of the water in just a few generations.

TERMS DEFINED: Dispersal, ploidy, diversification

ZEBRA FINCH

(Taeniopygia guttata)

TALKING THE WALK

Finches are the animal that started it all, back on Darwin's trip to the Galapagos, their beaks each perfectly attuned to their place in their ecosystem. But they also have a new story to tell.

BRAIN: has the capacity for "language"

BEAK: perfectly shaped for what it eats (holler, Darwin and guppies—page 45) but the song that comes out of it is the most recent surprise

RED COLORING: no gene for "red"

SEX DETERMINATION: more complicated than just birds and bees

WHEN IN (PRE)HISTORY/
WHERE ON THE RIVER

55 MILLION YEARS AGO
...after the K-Pg event, the great bird revolution brought about rapid evolution in most of the birds you see today.

7 MILLION YEARS AGO
...the group that includes songbirds seems to have made the largest genetic leaps. This idea of evolutionary or genetic speed is still a controversial one, but we're working with the knowledge we have, and changes are easy to see, plain as the beaks on Darwin's finches' faces.

A BIG STUDY HAS ONLY SCRATCHED THE SURFACE ON SOME CONNECTIONS AMONG THE NEOAVES, BUT IT DIVIDED THEM INTO THREE MAJOR GROUPS, OR CLADES:

- One with cuckoos and pigeons
- One with cranes and herons
- One including all waterbirds (minus waterfowl), be they diving, wading, or nearby nesting
- And one including nightjars, swifts, hummingbirds, and finches

This last group seems to be the fastest-evolving. According to a massive, cross-institutional, many-animal study in 2014, zebra finches are still never satisfied to stay the same. Even among the highly diverse new birds or neoaves (page 108), zebra finches have diversified more quickly than most.

K-PG EVENT

The K-Pg event (formerly the "K-T" event) is scientist shorthand for the Cretaceous-Paleogene mass extinction event, aka whatever it was that killed the dinosaurs. It happened around 65 million years ago and killed off 70% of all living things on the planet—maybe 80%, but it's hard to know for sure. Our current best guess is that the event entailed a massive impact from a meteorite or series of meteorites that would have raised so much debris into Earth's atmosphere that it blocked out sunlight for several months, killing green plants and disrupting the food chain.

BEASTLY BREAKDOWN

SMALL GENOME

One of those big changes that characterizes the bird genome is the size of the gene. All bird genomes are relatively small, about 1–1.26 billion base pairs on average. This means that in addition to making lots of changes, they're jettisoning the junk. More study will be required before we know how they're able to do that in the long run.

This iconic image of various Galapagos finches and their specialized beaks has come to symbolize Darwin's theory of natural selection. Ornithologist John Gould carefully illustrated the finches for Darwin's 1845 book *The Voyage of the Beagle*, **which predated** *The Origin of Species* **by fourteen years.**

SEX DETERMINATION

Sex determination is one of those phrases that simply describes what it is: sex is determined by the combination of chromosomes, Z or Y in the case of most songbirds, 20 of each. But we don't actually know how chromosomes (which are just large bundles of DNA) determine a bird's sex. It's not just a matter of male gene and female gene; there's an interplay between the two chromosomes, and there are a cascade of consequences that follow. So we're lucky that the zebra finch, our model species of the avian world, has not 40 chromosomes, but in fact 41. This extra chromosome is a genetic anomaly whose purpose has eluded researchers for decades, until technology advanced enough to unpack the gene arrived in May 2018. For now, the project can only tell us that the chromosome has indeed recently undergone evolution, which means that it's actively affecting the rest of the

bird's genome in some way. As researchers continue to dig, they hope this telltale chromosome might illuminate the mechanisms at play in the actual determination of sex determination.

THE COLOR RED

Picture the brilliant red in some animals' coats and feathers. In the animals that possess it, the red color is an advantage. In zebra finches, cardinals, and flamingoes, females prefer males with brighter reds (another example of sexual selection). Yet the origin of the red pigment isn't within the bird's bodies. That is, there is no red gene in birds. The color emerged over time, from an effect of eating carotene—the stuff that makes carrots orange. And yes, even humans can turn orange if you eat too much of it (it happened to my aunt as a child, shortly after my grandad brought home a free juicer and went a little overboard).

Confusingly, carotene is yellow, so some sort of conversion still had to happen in the birds' bodies in order to bring out the red pigments. Since red canaries became fashionable in the 1920s, scientists have sought out whatever enzyme or bodily process results in conversion. In 2018, a group of European researchers isolated this gene.

But in order for this color to matter, to affect success, other animals have to be able to see it. If the animals' bodies don't produce the color, why would their eyes be able to see it? It's not a chicken-and-egg question but a red and red-seeing-eye question: which came first?

In 2018, a group of British and Swedish researchers also found that this gene encodes for proteins in the eye. Any animal that can see the color red has this gene in their retina.

Likewise, that red-converter gene is in a greater family of genes, or genes that usually appear next to one another. These related genes have functions like controlling acid in stomach bile and controlling traits that look different in males and female animals (sexual dimorphism), like how a male cardinal is red and a female cardinal isn't.

The 2018 study draws this comparison, suggesting that the red-converting enzyme isn't about liking the color red. It's about what the color red communicates: "Hey baby, my offspring can handle digesting poison and will not die." Or maybe, if the female can see the red, it means their genes are compatible. Or they've both been tricked into perpetuating the genes by accident: the carotene-heavy plants benefit from the red because the more often they're eaten, the more likely they are to disperse their seeds (via animal poop or by riding along on feathers).

Over time, the potentially deadly action of eating a potentially poisonous new plant became connected to the reward of getting laid.

The answer to the chicken-and-egg red-and-red-seeing (and red-loving) question is that none of these came first. And they're no coincidence. Eating red, being red, seeing red, and loving red all evolved together.

THE "TALKING" GENE

In 2016, zebra finches again appeared at the center of a major revelation in evolutionary biology. Researchers in the US isolated a gene in zebra finches that seemed to correlate with its ability to make, remember, and complexify its song.

What's more, the gene had an uncanny analog to the FOX gene in humans, which is required for the development of speech and language. In the finch, the gene helps connect songs the bird is hearing with its physical ability to make the sound itself.

LONGEVITY (OR NOT)

Because of the finch's close relationship with sound, it's possible that too much sound actually shortens its life. A 2015 study found that once zebra finches left the nest, the ones that were more exposed to traffic noise lost telomeres faster than those in quieter environs. Telomeres are caps on the ends of chromosomes that protect genes from damage. Shortening of telomeres is one indication that an animal is, biologically, aging.

LESSON OF THE FINCH: The finch is a model organism for birds at large, which means that the fancy and elaborate traits highlighted here are just a cross section of the wonders you'd encounter venturing farther down the bird branch of the evolutionary river.

TERMS DEFINED: K-Pg event

GRAY SHORT-TAILED OPOSSUM

(Monodelphis domestica)

GREAT GRANDMAMMAL

By the time you're done reading this section, you'll be able to get know-it-all-y the next time someone at a party asks the actual difference between a possum and an opossum. But more importantly, you'll be able to get all know-it-all-y about how opossums are holding it down for old-school mammals and how having a baby the way humans do is actually a super weird freak accident of evolution.

WHISKERS: simple, compared to modern mammals

IMMUNE RESPONSE: simple, compared to modern mammals

SCALE 2–3 feet (61–91 centimeters, not including 8–13 inches [20–33 centimeters] of tail). Depending on where they live and what's on the menu at the time, they can weigh 4.6–13.2 pounds (2.1–6 kilograms).

FAM-O-METER

75%

of your genes line up with those in the opossum—maybe. And much more so in some areas than others.

TAIL: not really that prehensile; you're thinking of the *other* opossum (but not the possum, which is another thing entirely)

MILK: innovative for mammals, back in the day

SEX ORGANS: simple, compared to modern mammals

WHEN IN (PRE)HISTORY/ WHERE ON THE RIVER

Mammals didn't always gestate their babies inside their bodies, feed them via placenta and umbilical cord, and give birth many weeks or months later. Long before the modern-day placental-having mammals, there were metatherians, furry mammals that gave birth to very underdeveloped young and kept their little pink bodies in a pouch against their bodies until they were old enough to walk on their own.

165-180 MILLION YEARS AGO
...during the reign of dinosaurs and their contemporaries, another group of terrestrial animals diverged into two groups: metatherians and the group that would become placental mammals of today.

160 MILLION YEARS AGO
...the metatherians began to appear, diverging from a small, piddly group of what would become placental mammals.

150 MILLION YEARS AGO
...early mammals were already diversifying, surviving on whatever they could, filling a variety of niches. Early relatives of modern-day placental mammals were the eutharians.

75 MILLION YEARS AGO
...metatherians were having their heyday, at 68 known species in modern-day Europe, Asia, and North America.

65 MILLION YEARS AGO
...opossums probably split from the lineage that would become all other marsupials. They were early relatives to modern-day marsupials, including North and South American opossums, Australian possums, kangaroos, wombats, and—well pretty much most of the mammals in Australia and New Zealand.

55 MILLION YEARS AGO
...some species walked across the then connected Americas: northern animals southward, including most early metatherians.

BEASTLY BREAKDOWN

EYES

Opossums, like most nocturnal animals, have big eyes and enormous pupils. On the Australian possum, it's cute. On the Virginian opossum, it only adds to their terrifying visage.

MILK

The proto-opossum was really the first known animal to evolve two key phenomena that were important in the evolution of milk. One was the development of antimicrobial effects, hindering pathogens and keeping babies safe. The other, simultaneously, was probiotic effects favoring beneficial microorganisms. Though milk emerged much earlier than opossums (see platypus, page 20), this animal really perfected the clean milk process. These innovations made milk a winning adaptation, which allowed marsupials, and later placentals, to thrive.

SEX ORGANS

Even though it's a mammal, the female possum has simpler sex organs. The chute that it uses to push out its young is more like a bird's cloaca. This makes sense, because it's not so much pushing out whole babies as soft little eggs. There is no uterus or vagina.

IMMUNE RESPONSE

Opossums' and other marsupials' systems of immune response are simple. Genetically, they look much different than ours. Inflammation, for instance, is a key feature of immune response in mammals, but it wasn't so for proto-opossums; it's one of the genetic bits they're missing.

That means that inflammation came with the change, and that means that the very act of pregnancy as we know it may have started as an elaborate immune response.

That's weirder than coming from monkeys. Think about it: if not for some weird proto-opossum infection, we mammals of today might never have come about. Or maybe it's not so weird. Looking through this book, it's clear that disease avoidance (immunity and cancer control) have driven evolution more than food or sex. Kind of gross, but true. The Australian researcher behind the study showed that the complex mammalian immune system arose before the two lineages of mammal separated. She compared 1,500 immune genes in the opossum genome to those in the human genome and found a lot of similarities.

The moment when a fertilized egg latches on to the wall of a mother's womb to begin becoming a baby is called implantation. Implantation is, from the body's point of view, a lot like a parasite latching on to its host. Yet in between implantation and birth, the immune system is held in check, allowing the fetus to thrive.

Marsupials have very short pregnancies.

SAYING POSSUM

Opossum: South American. Many species including our gray short-tailed star. Family Didelphidae.

Possum: Only in Australia. Entirely separate family *Diprotodontia*. Also marsupials, but possums have, for instance, fluffier tails.

Also opossum: North American. Just a single species left: Virginia opossum. Family Didelphidae. Americans refers to them colloquially as possums.

Playing (O)possum

The gray short-tailed opossum is a very different beast than the Virginian opossum we're used to seeing in North America. The Virginian opossum is most often depicted as hanging upside down, carrying things with its tail, and playing dead.

Now *this* is a Virginia opossum (*Didelphis virginiana*). The 1897 illustrator for the German encyclopedia *Meyers Lexikon* probably never saw one in real life, accounting for its overly short and mouse-like facial features.

Early opossum embryos develop for about 12 days, enclosed as shelled eggs in the womb. They then shed their shells and try to attach to the uterine wall, activating placenta-promoting genes. But after about two days, the mother's immune system "rejects" the embryos, causing the birth of a litter still at a very immature developmental stage compared with placental mammals.

But they don't have much immunity. Newborn opossums have extremely simple and weak immune systems, a common trait among marsupials. But they *can* regenerate a severed spinal cord. This trait makes them perfect candidates for medical research on organ transplantation and cancer.

TAIL

Opossums have prehensile tails, meaning that they have motor control over their tails to the end, allowing them to grasp tree limbs—but not as much as pop culture would have you believe. Although they are often pictured hanging upside down by their tails, it's a myth that they do so while they sleep.

WHISKERS

A recent study showed that opossums have similar but not identical whisker function to their placental distant cousins. They *do* similarly have a single whisker that functions as the so-called best whisker, or BW (because scientists must abbrev e-thing), that leads the way for all the other whiskers. But the corresponding BW response in the possum's nervous system doesn't seem to be as strong as in mice and rats, who have specially evolved whisker-to-brain pathways.

TRANSPOSONS

One clue to the differences between opossum and human is that their genome lacks a number of non-coding "junk" elements that are conserved in placental mammals. That means that these elements must have helped mammals

THROUGH DARWIN'S EYES

One of the ways Wallace was most helpful to Darwin was in breaking down the geographic aspects of evolution, outlining his theories of migration among the species. His account of the intercontinental journey of the opossum is an excellent example of this:

We have here the most remarkable case, of an extensive and highly varied order being confined to one very limited area on the earth's surface, the only exception being the opossums in America. It has been already shown that these are comparatively recent immigrants, which have survived in that country long after they disappeared in Europe. As, however, no other form but that of the Didelphyidæ occurs there during the Tertiary period, we must suppose that it was at a far more remote epoch that the ancestral forms of all the other Marsupials entered Australia; and the curious little mammals of the Oolite and Trias, offer valuable indications as to the time when this really took place.

become what they are and evolve into their current, "essential" form after proto-all-the-mammals split from the proto-opossums. Researchers estimate that 95% of the innovations in the genomes of placental mammals came after the split.

LESSON OF THE OPOSSUM:

The secret to understanding the evolution of the placenta lies in the wet pink pouch of a mammal that's gone millennia just fine without one.

TERMS DEFINED: Marsupial, placental

NINE-BANDED ARMADILLO
(Dasypus novemcinctus)

COUSINS FROM WAY BACK

They might not look much like each other now, but these animals are relatively close cousins. Much like the horse, each of these has an ancient ancestor that shows up in the fossil record as long as 35 million years ago, already looking a lot like the animal it eventually became.

SCALE
15–23 inches
(38–58.5 centimeters, without tail)

BABY TEETH: nine-banded armadillos lose their baby teeth and get adult ones, like other mammals, but not like other Xenarthrans.

DEFENSES: taking it as far as is logical and then some, with a covering like a tortoise*

FEET: walks on tips of claws

(*not related to a tortoise)

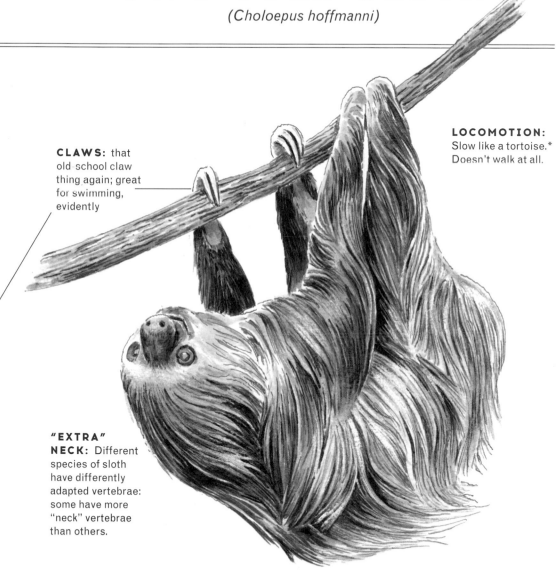

HOFFMAN'S TWO-TOED SLOTH

(Choloepus hoffmanni)

CLAWS: that old-school claw thing again; great for swimming, evidently

LOCOMOTION: Slow like a tortoise.* Doesn't walk at all.

"EXTRA" NECK: Different species of sloth have differently adapted vertebrae: some have more "neck" vertebrae than others.

SCALE
21–29 inches
(53.3–73.6 centimeters)

FAM-O-METER
70%?

These cousins have evolved to be different from humans in two very different ways. The human vs. sloth/armadillo genome comparison isn't a high priority, but the results could be interesting. For instance: What's with the fact that humans and armadillos are the only two animals to carry the same strain of leprosy? Studies say we probably gave it to them, and their unique body temp allowed it to thrive. But what do the genes say?

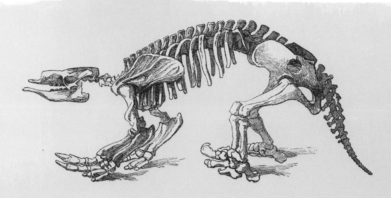

THROUGH DARWIN'S EYES

It's no coincidence that some of the most important fossil finds to date were made by Charles Darwin himself. Being able to observe the difference between ancient and modern iterations of animals is one of the privileges that he credits to helping him cement his theory. In particular, he credits large mammal fossils including the giant ground sloth in South America. "Fossil bones," he called them. He wrote his sister Caroline that the best finds could "tell their story of place and time with almost a living tongue."

The first person to discover a fossil of the giant ground sloth known as *Megatherium americanum* was a Dominican friar named Manuel Torres. His discovery became the first prehistoric mammal skeleton ever to be displayed in public. It was so complete that French anatomist Georges Cuvier was able to describe it accurately from a drawing, having never seen the actual specimen. This detailed description was among the reading material Darwin had with him on the *Beagle*. So, when he found his own specimen, he knew what he was looking at. Unlike modern sloths, *Megatherium* probably weighed in between 4 and 6 metric tons, and stood 13 feet (4 meters) tall. And I mean stood: fossilized *Megatherium* footprints show that it stood and even walked on its hind legs.

Also near and around Buenos Aires, in 1832, Darwin found pieces of what he described in his notes as "an armadillo on a grand scale." He correctly guessed that it belonged to a 10-foot-long (3 meters) prehistoric cousin of the sloth and anteater called a glyptodont. A year later, in 1833, a guide led him to a semi-unearthed fossil of a nearly complete shell, and he was never more certain of his find. He said in a letter, "Immediately...I thought they must belong to a giant armadillo, living species of which genus are so abundant here." But before he ever published his findings, he read that the famed naturalist Georges Cuvier had been insisting that these types of bony bowls in fact belonged to the giant ground sloth, *Megatherium*. Darwin respectfully toed this line for years afterward, even identifying his findings as *Megatherium* remains in the 1844 essay that was his first major publication about his ideas about evolution.

WHEN IN (PRE)HISTORY/ WHERE ON THE RIVER

About 100 million years ago, after platypuses and echidnae and all the crazy marsupials of Australia went their own way, the next group to split off was the group that would become elephants, mammoths, manatees, and armadillos, sloths, and anteaters. Pretty immediately after, ancestors of what would become sloths/armadillos/giant anteaters headed to what is now Central and South America, while the proto-elephant/mammoth headed to Africa. Living members of this former group are collectively called xenarthrans, which sounds badass. The entire group includes 20 species of armadillos (including one called the screaming hairy armadillo), six species of sloth, and five species of anteater.

Two-toed sloth from the 1821 Die/Das Skelete series by engraver and naturalist Joseph Wilhelm Eduard d'Alton.

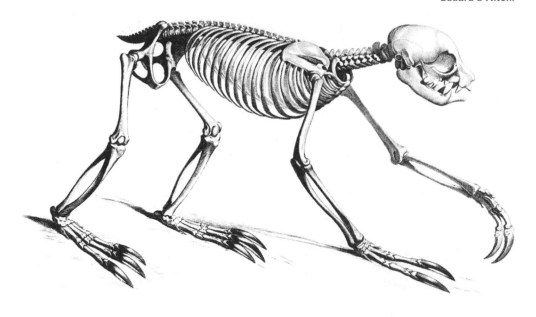

COMPARATIVE BEASTLY BREAKDOWN

CLAWS

The long, strong claws of both the armadillo and sloth appear much like the claws of their prehistoric ancestors. When the armadillo isn't using its claws to dig burrows or seek grubs, it has to walk around tiptoe on the tips of them. The sloth is also terrible at walking on its claws, and instead uses them to hang in trees.

DEFENSES

Armadillos have evolved a tough shell or carapace, covered in hard, scale-like plates called scutes—the same skin covering as makes up a turtle's shell, the lumps on the crocodile skin and bird feet, and the "scales" on sturgeon. Scutes are made of keratin and held in place by thick, leathery skin. When confronted with danger, they shut themselves up in a ball, their head plate and tail plate perfectly aligned to close and seal them tight.

In contrast, sloths' best defense is simply moving so slowly that predators are unlikely to notice them. They have some of the slowest myosin in the animal kingdom—they couldn't move more quickly if they wanted to.

Vintage European illustration of armadillos in Texas; year and artist unclear. But check out those cute scutes.

NOT XENARTHRANS

For centuries, naturalists included in this group aardvarks and the gorgeous, plate-armored pangolin. It was a reasonable assumption: they share body shape, right down to their long snouts and tongues and digging claws, custom-made for a diet of termites and ants. But thanks to genomics, we now know that these similarities are just a classic case of convergent evolution: similar lifestyles led to similar body shapes. Both animals have been reclassified to a taxonomic order all by themselves. Aardvarks are still a bit of a mystery, but they seem to be most closely related to elephant shrews and a weird mole-like creature called the tenrec. Pangolins are most closely related to carnivores like cats and dogs.

CIRCADIAN RHYTHMS

These unusual defenses subsequently caused their owners to evolve unique aspects to their body clocks. Armadillos' shells are so thick that it makes it hard for them to regulate their body temperatures, so they have to change their habits to match the season: they're largely nocturnal in the summer but will switch over to daytime activity in the winter months to maximize sun time, while escaping to their cozy burrows at night.

Meanwhile, sloths avoid would-be predators by maximizing their time safely in the treetops. And because the only time they regularly come down out of the trees is to evacuate their bowels, they've evolved a slow digestive tract. Digestion, we now know, is another function controlled by genetic circadian rhythms, and sloths go a whopping seven days between descents. They've also evolved enormous large intestines, affording them what is possibly the largest single-poop-to-body-size-ratio in the vertebrate family tree.

LESSON OF THE ARMADILLO/ SLOTH: Some animals start out weird, wind up weird, and keep it weird in between. Let's see if we can't replace the phrase "survival of the fittest" with "keep animals weird."

SEX ORGANS:
making babies
really really fast

INTERNAL ORGANS:
a lot like humans only really
really fast

MUSCLES:
really really fast

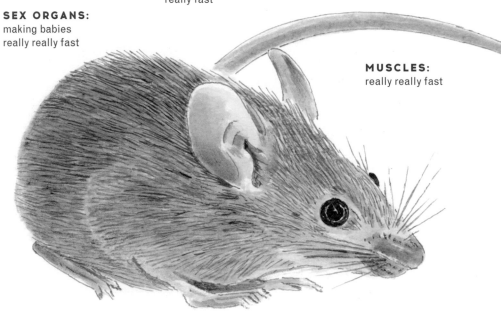

SCALE
3.0–3.9 inches
(7.5–10 centime-
ters), plus a
2.0–3.9 inch (5–10
centimeter) tail

FAM-O-METER
85%–
99%

It depends on how you count it. Records of this number have changed over the years. The earliest human/ mouse genome comparison claimed 99% overlap. As researchers published increasingly complete genomes, the percentage slipped downward. But now, longer-term, in-depth comparisons are showing that genes that seem superficially different in mice and men may turn out to be more similar than we could have imagined.

HOUSE MOUSE

(Mus musculus)

OF MICE AND MEN

When it comes to genetics, mice just might be our closest allies and most important distant cousins. Imagine a lab animal that, like the African clawed frog, fruit fly, or zebrafish, can multiply quickly and have a lot of young—that's the house mouse. The place at which the mouse and human lineages diverged was basically this: either you're an animal that reproduces a lot, or you're not. Mice went that direction, and the primate lineage went another. But mice and men still share enough in common that this section will make you think twice before you set another mousetrap.

WHEN IN (PRE)HISTORY/
WHERE ON THE RIVER

Believe it or not (though by now you probably do believe it), mice and rats are just a few major branches from humans on the mammalian river. Their lineage broke from ours about 75 million years ago.

Rodents are our closest non-primate relatives, but more importantly they've retained much of the same gene families that we have. This means that the major goings-on in a mouse are the same major goings-on in us.

Think about it this way: the average mouse genome might have 85% overlap with the average human genome. But for many of the lab mice in medical testing facilities, that percentage is much higher; they've been injected with additional human DNA for testing purposes, and it doesn't kill them. Likewise, when medical facilities make findings about humans via experiments on mice, we better hope it doesn't kill us.

BEASTLY BREAKDOWN

GENE INTERACTION

There are as many ways to perform experiments on mice as there are medical labs in the world. We inject them with our cells, cancers, chemicals, and mutations; slice up their DNA; and see what happens. For example, in 1991 an Italian lab grew a naked mouse with a human ear growing out of its back, by injecting ear-specific stem cells under its skin.

SEX DIFFERENCES

Mice, like humans, have XX and XY chromosomes. The arrangement of chromosomes determines what sex an animal becomes, but it can also affect a lot of other things. In fact, it can be life or death.

In 2017, a researcher named Joanna Floros published findings that really hammered home the importance of paying attention to sex differences when treating major diseases. She conducted a crucial study across male and female mice that had been spayed or neutered and those that hadn't. She was looking to see if there was a sex division between animals that were exposed to the exact same DNA manipulation that caused them to develop lung cancer. She learned that female mice developed the cancer more often than neutered male mice, but the results were nil in the control group. She'd proved how much sex differences matter, but not why. The answer may be found in our micro-RNA, which is the next frontier in DNA research.

MICRO—RNA

We've learned about transposons: bits of genetic information that move around to make changes in a genome. Micro-RNA (mRNA) are transposons that are so fragmented they don't even seem to be useful for anything. Still, though, micro-RNA is RNA—genetic material. And we know by now that any genetic material is good genetic material. Right? Time, and more research, will tell.

SEX ORGANS

One of the major differences between mice and men, however, since we split, is in reproduction. Their newer genes disproportionately affect the reproductive system, making them able to have more babies, faster, and more often.

MUSCLES AND MYOSIN

Another major change between mice and men went into coding for muscles. Their muscles have gotten smaller and myosin has gotten faster.

BLOOD

Mice genes show that their blood is more efficient at processing oxygen than other animals in its lineage.

Mice from Valparaíso, Chile, drawn by Captain FitzRoy of the *HMS Beagle*. Darwin originally dubbed them *Mus longicaudatus*, but they have since been reclassified as *Oligoryzomys longicaudatus* or rice rats: rodents more closely related to hamsters, voles, and lemmings than to the old-world house mouse.

"HUMAN EAR"

This clever play on words is meant to reference the mouse with the human ear growing on its back, but has nothing to do with actual ears so much as what goes into them. A 2009 study took a look at what a mouse might do if it had genes for human language. Remember that series of FOX genes found in the zebra finch? The ones that, in songbirds, code for learning and sharing vocal communication? The same gene that helped lead to the development of human speech? Mice have that, too. We're just not yet sure what they're doing with it.

LESSON OF THE MOUSE: No matter how you feel about animal testing, and no matter how you feel about where humans came from, the genetic similarities between human and mouse are close enough that research conducted therein has saved thousands of human lives. This is a fact, and that fact is a product of evolution.

At the same time, a mouse isn't exactly the same as a human any more than one human is the same as another human—or even a male is to a female of any species. This is a fact, too, one that encourages us to keep researching the evolutionary connection between mouse and man.

TERMS DEFINED: Micro-RNA

BRAINS: slightly more cooperative than chimps

HANDS: can make and use tools, communicate

BODY SHAPE: slightly more like humans than chimps

FIBULARIS TERTIUS: tiny leg muscle previously thought to appear only in humans

SCALE
45 inches (115 centimeters) standing

FAM-O-METER
98.7%

Exactly tied with chimpanzees for our closest (still very distant) cousin (many times removed).

BONOBO
(Pan paniscus)

OTHER BROTHER FROM ANOTHER MOTHER

That quippy headline was too good to pass up, but I really can't emphasize often enough that any close family relationship you can name—mother, father, sibling, granddaddy—none of these is a good analogy for the relationship between humans and apes. They have their own family thing going on.

WHEN IN (PRE)HISTORY/ WHERE ON THE RIVER

Human researchers from Germany sequenced the bonobo genome back in 2012, and found that bonobos and chimps themselves share 99.6% of their DNA (0.3% less than we share with Neanderthals, Denisovans, and other early humanoids).

Curiously, though, about 1.6% of the 98.7% overlap we share with the bonobo, we share only with them, but not with chimpanzees. And the same is true the other way around.

These numbers tell a story. Our common ancestor, the ape-like primate that had yet to evolve into either modern ape or modern human—well actually, it's more accurate to describe that animal as a whole population. Each individual in a "species" doesn't have the exact same genes, after all. This ancestral population was quite large and genetically diverse. Genome-crunchy stats suggest that there would've been 27,000 breeding individuals in this group. Once the ancestors of humans split from the ancestor of bonobos and chimps more than 4 million years ago, the common ancestor of bonobos and chimps retained this diversity until their population completely split into two groups 1 million years ago. The subsequent offspring of these groups would eventually evolve into bonobos, chimps, and humans. Each of our groups would retain slightly different permutations of that ancestral group's diverse gene pool.

While we take a moment for that idea to set in, I'll point out that this scenario is why I prefer my river analogy for evolution, as it has

An illustration of a chimpanzee from Wallace's 1889 book Darwinism: *An Exposition of the Theory of Natural Selection, with Some of its Applications.*

room enough in it to envision the collective flow, rather than a single-file line. Each lineage doesn't have a single missing link; it has millions. Evolution always, necessarily happens en masse, with breadth as well as length. (Unless maybe you're a whiptail lizard.)

The common ancestor of humans split with the common ancestor of bonobos and chimps about 4–7 million years ago. Chimp and bonobo populations split about 1 million years ago, and they haven't interbred since. They probably could if they wound up in the same place as each other, but the two groups have been geographically separated by the Congo River for at least that long.

So what exactly are those small differences in chimp and bonobo all about? How, in all that time, have they grown apart? We're not sure yet, but we have some leads.

BEASTLY BREAKDOWN

IMMUNE RESPONSE

There's a version of this gene thought to help fight retroviruses that show up only in chimps. In fact, chimps get a milder strain of HIV than we do (called SIV—simian immunodeficiency virus). Bonobos, however, aren't susceptible to any strains of SIV yet found. In 2017, American researcher Emily Wroblewski compared the immune-related genes in bonobos and chimps suffering from SIV. She found that the populations of bonobo all shared three variations of a certain gene that codes a protein that helps immune cells recognize viruses as viruses—the exact part of the cell that immunodeficiency viruses attack. This gene is shared by bonobos, chimps, gorillas, and humans, meaning the gene itself came from our shared ancestor. But only bonobos have enough variation in that gene to have kept them safe so far. Once again, variety isn't just the spice of life; it's the key to it.

BODY SHAPE

Bonobos tend to be slightly taller and less muscular than chimpanzees.

TELLTALE MUSCLES

The fibularis tertius is a strange little muscle found in the lower leg of humans but not in chimpanzees. For decades, this muscle was touted as a trait of man alone, and some theorists even went so far as to cite the muscle as the single most important innovation to full bipedalism (walking on two legs). But in May 2018, an anatomist named Rui Diogo found the muscle in a small number of bonobos. Of the seven bonobos he dissected, three had the muscle, appearing just as it does in humans. And Diogo didn't just dissect bonobos. The entire project entailed dissecting several specimens of a variety of great apes, specifically hunting for seven different muscles and tendons that science had theretofore claimed to be "human-specific." He found all seven in one or more of his bonobo, chimp, and gorilla specimens. This included two muscles thought to have been associated with vocal speech, one in the larynx, and one in the face next to the jaw.

SEXUAL PROCLIVITIES

Differences among hormone-related areas of the bonobo and chimp genomes may one day explain why bonobo females are less sexual during menses—though we're not sure what that means for us.

SOCIAL INTELLIGENCE

Another difference could lie in the way we perceive social cues. Bonobos tend to be more sharing and cooperative, while chimpanzees tend to be better with tools and spatial reasoning but more aggressive with each other.

One area of the bonobo genome is possibly tied to understanding social cues. Bonobos and humans have it but chimps don't. But we're not talking about social subtleties here. This area is deep-genome stuff going back to urine-related behaviors. Chimps don't respect urine marking, whereas bonobos do. We're definitely not sure how that manifests itself in modern humans—it's too early in the study to understand what the pee-language analog might be. And I won't go so far as to say, "Chimps would reach for the pee joke here, but bonobos and I are more evolved." We're just differently evolved.

LESSON OF THE BONOBO:

The subtle but significant differences between chimps and bonobos speaks volumes about the subtle but significant changes that occur in even small sections of DNA. (No accounting for taste, but bonobos kinda seem like a better version of chimpanzees. Would a few tweaks to human DNA give us a whole better species of human?)

PART 3:
HOPELESS MONSTERS

Inspired by turtles, a mid-century naturalist and early geneticist named Richard Goldschmidt coined the term "hopeful monster" as a way of describing organisms of extra-evolutionary weirdness. Like the turtle, with a shell made from a severely malformed spine, these organisms were ones that appeared to have had what he called a "macromutation"— a large-scale mutation like an extra limb or horribly malformed underbite, which appears to be something that would get the organism killed straightaway. Though the term might be more poetic than useful (#sciencebandnames), it refers to animals that possess a nonsensical or "monstrous" adaptation that seems more trouble than it's worth but ends up driving a huge new surge in evolution, giving way to many more successful species of its kind.

Goldschmidt included in this category animals that he saw as standing alone in the animal family tree: the turtle, the platypus, the giraffe, and even humans with their big special brains. We now know that the reasons for these animals' unique features isn't due to a single mutation but rather a complex and fine-tuned calculus of their genomes. Goldschmidt was onto something with the notion that mutations happen by accident and that their utility to the organism either plays out over the course of its life and allows it to reproduce, or it doesn't.

This chapter takes another look at those apparent outliers at the end of their evolutionary rivulets and examines how they came to be there and what their monstrous features have to do with it.

BEAK: a facial bonus formed by a cluster of bonus genes that reptiles and mammals don't have

COMB: a focal point of sexual selection (and red—see zebra finch, page 168)

FORELIMB: about as useful as a *T. rex*'s

FEET: covered in scutes like most crocodiles, turtle shells, and armadillos

SCALE
28 inches (71 centimeters). About the size of a chicken.

FAM-O-METER

70% of our DNA overlaps with a chicken's. They don't have as many base pairs as we do, but they do have about the same number of complete protein-coding genes.

JUNGLE FOWL, AKA DOMESTIC CHICKEN

(Gallus gallus)

WHICH CAME FIRST? THE CHICKEN

Jurassic Park has already arrived, only the dinosaurs are fat, full of hormones, and deep-fried by the bucketful.

WHEN IN (PRE)HISTORY/ WHERE ON THE RIVER

Today, chickens are ubiquitous. There are more chickens on Earth than there are humans. Say what you will to vegetarians, but that technically makes them a more successful species than us.

Chickens, turkeys, and, believe it or not, ducks and geese are more like dinosaurs than other birds (a revelation from a 4-year "big bird bang" chunk of studies that sequenced 48 species). This is an analysis based both anatomical comparisons and phylogenetic analysis of living birds, which tells us when that fowl broke off on their own.

250	MILLION YEARS AGO	65	0
MESOZOIC		CENOZOIC	

...Around 103 million years ago, before the "bird big bang," some of these proto-birds split to go their own way. The first of these was the group that would become the flightless ratites: ostriches and emus and kiwis.

CHICKEN

...Around 89 million years ago, the proto-chicken/turkey/duck lineage bounced off in its own direction.

BEASTLY BREAKDOWN

FORELIMB

In the 1980s, *Drosophila* researchers discovered a gene that controlled whether or not a feature would show up within a certain body segment. In the fly, the gene controlled the bristles on a fly's butt, so researchers named it hedgehog. But it was in chickens that researchers connected the hedgehog to the development of forelimbs. In vertebrates, it's the hedgehog gene that helps the body decide if the cells for the forelimb are going to become a wing, a human hand with a finger, a bat wing, or a flipper.

Chickens are great for evolutionary developmental research because their embryos are so easy to access—in eggs, of course! Thus,

researchers figured out how the hedgehog gene was working in chickens by dyeing molecules in chicken embryos and watching where they activated using an egg light.

BEAK

In 2015, armed with the knowledge that chickens and dinosaurs shared an ancestor with the alligator, geneticists spliced genes so that an alligator's snout grew onto the beak area of a chicken. The goal was to see how a beak evolved, and the embryo looked somewhat like a velociraptor. Jurassic Park is not just a movie, people.

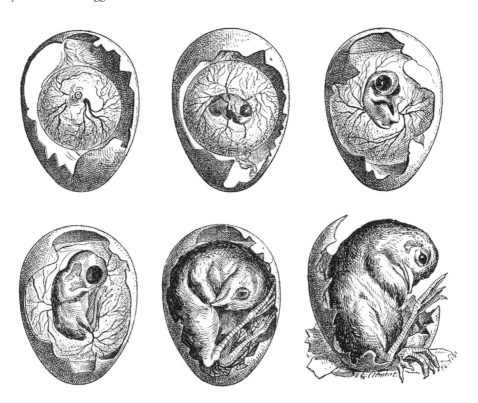

An 1895 engraving titled *Development of a Chicken Egg*, from *Dictionnaire des mots et des choses* (Dictionary of Words and Things) by grammarian duo "Larive et Fleury" aka Auguste Merlette and Aimé Hauvion, respectively.

LESSON OF THE CHICKEN: We've spent a lot of this book insisting that success is simply defined by persistence and actually has nothing to do with being better than or winning. But then I think of the chicken. It's been around longer than most of the birds on the planet, and it has us breeding it for lovely plumage, filling it with feed until it's fat, and giving it bionic upgrades to look like the long-lost dino-ish relative still alive and well in its genome. And I think, *Maybe we can have it all.*

THROUGH DARWIN'S EYES

In his writings and talks, Darwin insisted that the domesticated chicken arose from the red jungle fowl. Genomic evidence, though, shows that the domestic chicken is more related to the gray jungle fowl. His descriptions of both are boring, however, so here he is delightedly describing a different jungle fowl altogether: "I also obtained here a specimen of the rare green jungle-fowl (*Gallus furcatus*), whose back and neck are beautifully scaled with bronzy feathers, and whose smooth-edged oval comb is of a violet purple colour."

GHARIAL
(Gavialis gangeticus)

BIRD BRAIN

Crocodiles used to be the template for dinosaur. Artist renderings of those prehistoric "thunder lizards" had croc-like faces and skin, and the crocs that walk among us were considered "living fossils." Today, birds have stolen the crocodile's dinosaur thunder, and we know the term "living fossil" is a misnomer: Nothing living is set in stone. But crocodiles like the Gharial have been around for a long time. They have *looked* about the same for all that time. They're living proof that evolution marches on...but sometimes it's a slow march.

LEGS: smooth but weak like a lungfish's

WHEN IN (PRE)HISTORY/
WHERE ON THE RIVER

By now you're familiar with the concept of common ancestors. And one of the coolest relatednesses is that common ancestor who would have connected the archosaurs: a proto croc/bird/dinosaur thought to have lived around 275 million years ago.

In 2014 an international genomic team sequenced the genomes of three crocodilian species: the American alligator, the saltwater crocodile, and the Indian gharial. They learned that the rate of molecular evolution in the crocodilians is far slower than in mammals. The most likely reason for this relates to the relatively long time between generations in crocodilians. This in turn may have had to do with the apparent fact that crocodile and gharial populations experienced sharp declines during the most recent ice age, only about 11 million years ago. Today, there are just 25 distinct species of crocodilian, none of whom are in a big hurry to adapt.

Taxonomists accounted for the gharial's unique physical characteristics by assuming it had split from other crocodilians early in their lineage, long before alligators and crocodiles began to diversify among themselves.

SCALE
11–14.5 feet
(3.3–4.4 meters)
for females,
16–19.5 feet (4.9–6
meters) for males

FAM-O-METER

69%?

Close to a chicken, but not quite a chicken.

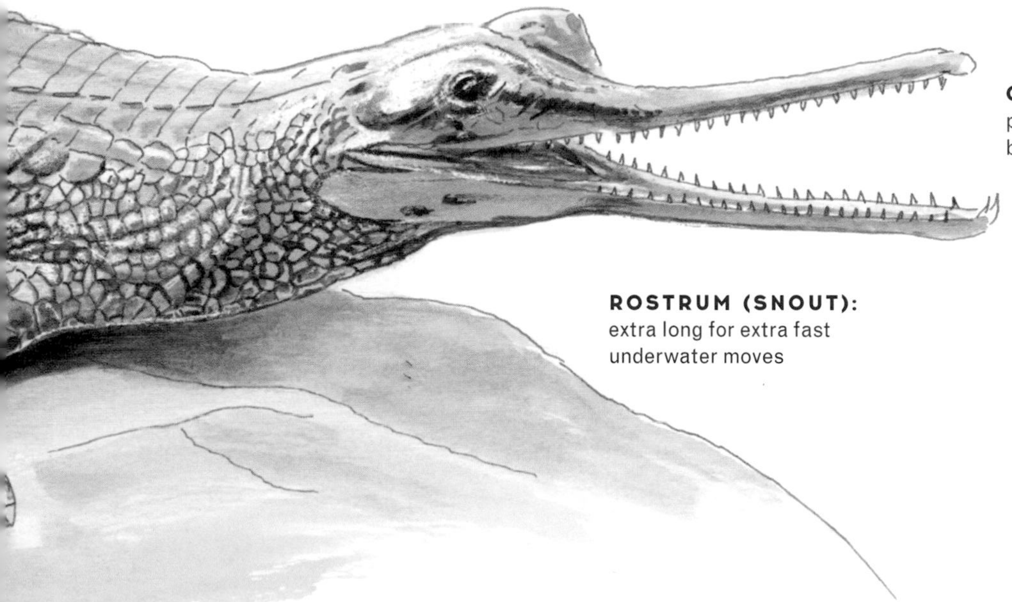

EYES: worn way up top to stay as submerged as possible

GHARA: nose pot for sexy bubble blowing

ROSTRUM (SNOUT): extra long for extra fast underwater moves

250-220 MILLION YEARS AGO

...the lineage that would become modern crocodilians split from the rest of the archosaurs.

137 MILLION YEARS AGO

...the lineage that would become alligators and caimans split off, further splitting into caiman and alligator around 98 million years ago. Alligators began to diversify around 89 million years ago.

34 MILLION YEARS AGO

...the gharial split from the false gharial, a smaller crocodilian previously given the scientific name *Tomistoma schlegelii* and placed within the crocodile family (instead of the alligator or gharial family). Genome studies in 2005 and 2007 evidenced closer relatedness between the two than previously thought, and recommended the false gharial be reclassified.

BEASTLY BREAKDOWN

GHARA

Gharials get their name from the ghara at the tip of their rostrum (snout), which is in turn named for the Hindi word meaning "clay pot." The ghara is a round knob that begins to grow over the nostrils of males at the onset of puberty, a full ten years into life. When the male gharial is ready to mate, he begins to exhale rapidly, which creates a buzzing against the ghara, which in turn amplifies it. The larger the ghara, the louder the noise, and the more likely the male is to attract a mate: sexual selection in action. If

Scientific record of the gharial's actual ghara were rare before the photographic era, even in paintings from India. This 1887 image from the *Meyers Lexikon* encyclopedia is actually one of the better ones in circulation.

the male's display is successful, copulation goes down in the water, where the exhaling turns into the blowing of a lot of bubbles. That is to say, some researchers have noted the bubble blowing as part of the mating ritual, but the logical conclusion suggests that the bubble blowing is just buzzing that has become inadvertently submerged.

ROSTRUM

The gharial subsists almost entirely on fish and large waterborne prey. Its distinctive snout is an adaptation especially good for this endeavor: it's long and thin and able to quickly clamp shut with minimal water resistance. Its 100+ long interlocking teeth instantly create a cage. Throughout a gharial's life, its snout continues to grow increasingly longer and thinner, maximizing its hydrodynamicity (which is like aerodynamic, only in the water). In fact younger gharials, with broader rostrums, tend to eat other, less fully aquatic prey, like insects, crustaceans, and frogs.

GIZZARD

Studies of gharials' stomach contents tell us that they're opportunistic eaters and sometimes carrion scavengers, which means sometimes they bite off more than they can chew. Like their avian cousins, gharials swallow rocks and hard objects that help grind up food in their stomachs. Interestingly, these hard objects sometimes include the jewelry of humans, whose funeral traditions include sending bejeweled human remains down the Ganges River.

EYES

Wide-set atop its head, the gharial's eyes are ideal for lurking in water and stalking fishy prey. Its pupils are vertical, like a cat's, a shape ideal for hunting.

FEET

Its feet are webbed. The gharial is at its most deft in the water.

LEGS

Unlike gators and crocs, which often venture out of the water to snap down on some red meat or fowl, the gharial is next to useless on land. Its legs are so short and weak that they scarcely lift their bellies off the ground.

SCALES

Whereas most alligators and crocs have bumpy, rough scales, a gharial's are smooth.

SEX DETERMINATION

Like the American alligator, the gharial has 16 chromosomes. Unlike organisms with genetic sex-determination systems (XY, etc.), crocodilians seem to not have sex chromosomes. Instead, sex is determined by incubation temperature of the egg—same as with turtles and tortoises.

LONGEVITY?

There is little existing data to help us understand the potential age of a wild gharial, but it is estimated to live as long as 60 years—good but not spectacular for an animal able to weigh in at up to 1,500 pounds. For comparison, male saltwater alligators, Earth's largest known extant reptiles, can weigh up to 2,000 pounds, reach sexual maturity around age 16, and probably live to be around 70 in the wild.

LESSON OF THE GHARIAL:

Never forget that with a single twist of fate, this big lumpy creature could be covered in feathers and flying south for the winter. Instead, it's been pretty much doing the same thing for the past several million years.

BARN OWL

(*Tyto alba*)

SURROUND SOUND WITHOUT A SOUND

The barn owl is uniquely adapted to its environment and lifestyle, and possesses characteristics no other bird has. Sometimes being a perfectly adapted, well-oiled, fine-tuned, pristine product of your evolutionary niche means having gangly legs, big ears (yes, bird ears), and a lopsided skull.

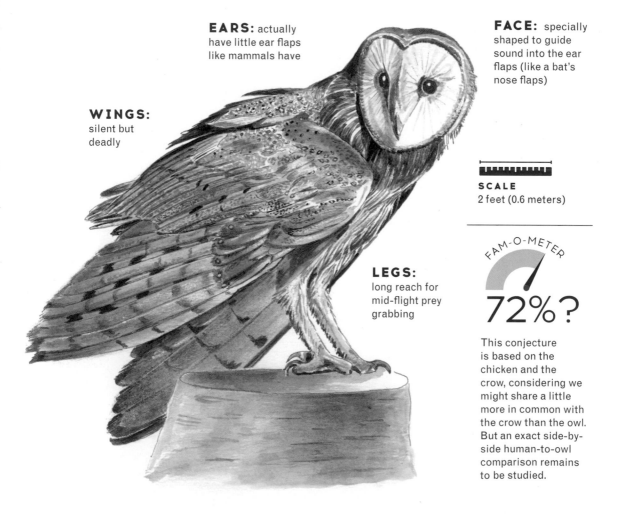

EARS: actually have little ear flaps like mammals have

FACE: specially shaped to guide sound into the ear flaps (like a bat's nose flaps)

WINGS: silent but deadly

LEGS: long reach for mid-flight prey grabbing

SCALE
2 feet (0.6 meters)

FAM-O-METER

72%?

This conjecture is based on the chicken and the crow, considering we might share a little more in common with the crow than the owl. But an exact side-by-side human-to-owl comparison remains to be studied.

WHEN IN (PRE)HISTORY/ WHERE ON THE RIVER

Owls are a raptor—a bird of prey, in the same general evolutionary branch as eagles and vultures. The exact number of owl species varies according to your source, but it's over 100 and most are in a genus informally known as "true owls" (screech owls, spotted owls, horned owls). About 18 owl species are in the genus *Tytoindae*, informally known as "barn owls," all closely related to *the* barn owl, *Tyto alba*. Barn owls are one of the most widely distributed groups of birds, thriving on every continent except Antarctica. *Tyto alba* was all over the Northern Hemisphere by about 25 million years ago.

BEASTLY BREAKDOWN

FACE SHAPE

Like other nocturnal animals, barn owls have large eyes, but not the largest. In fact, the barn owl appears to have smaller eyes than many owls because of the special shape of the feathers around its eyes and face. The signature heart shape of the barn owl's face comes from the way its facial feathers are arranged; they are uniquely positioned to create a sort of amplification hood to further boost its hearing accuracy.

SKULL

Everything about the barn owl's skull is asymmetrical. Its uniquely asymmetrical eye sockets and ear holes (with amplifying protrusions called ear flaps) help it triangulate sights and sounds not only with spatial precision but with precise timing as well.

EARS

Unlike some owls, barn owls can home in on prey using hearing alone. They can detect minute interaural time differences, and their hearing range is up to 10 kHz.

BRAIN

Barn owls have what appear to be the largest auditory processing centers among bird species.

WINGS

The barn owl's feathers have evolved to be utterly silent as they fly. They achieve this effect thanks to extremely soft, delicate feathers, which they never touch with preening oil or dust as other birds do to keep off insects or parasites.

TALONS

The barn owl's long legs and sharp talons allow it to kill its prey with the swift motion of a snatch, without stopping to crush it, for a smoother flight.

LESSON OF THE OWL: If you're going to be something, be the best.

VELVET SPIDER

(Eresus kollari)

THE ELEMENTS OF STYLE

There's never been a lot of funding available for spider research. They're not closely related to most of the other animals whose genomes have already been sequenced. To date, only four spider genomes have been even partially deciphered. That makes it difficult to use those references to piece together and analyze spider genomes.

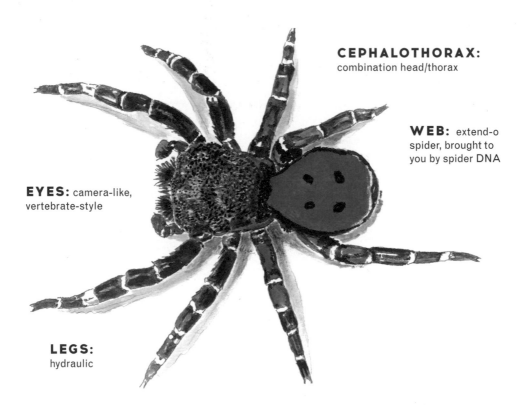

CEPHALOTHORAX: combination head/thorax

WEB: extend-o spider, brought to you by spider DNA

EYES: camera-like, vertebrate-style

LEGS: hydraulic

SCALE
Ladybird spiders (a type of velvet spider, male pictured above), are 9-16 mm (.35–.62 inch). African social spiders (*Stedodyphus mimosarum*), a different genus of velvet spider cited heavily in this section, are 8–14 mm (.31–.43 inch).

FAM-O-METER

60%?

This comparison, when it's run, is going to be fascinating.

But what we do know is fascinating. Mounting evidence even suggests that spider genomes in particular change dramatically according to their external environment (like the whiptail; see also octopus, page 255) and internal environment or microbiome (see mosquitoes, page 110; and humans, page 70). This might have something to do with the fact that the spider is one of the few animals whose body actually becomes part of its environment. Genomically and literally speaking, the spider has woven a tangled web.

WHEN IN (PRE)HISTORY/ WHERE ON THE RIVER

There are over 35,000 species of spider world-wide, ranging from massive, 4.5-foot-wide (1.4 meters) species in the tropics to tiny, lonely spiders in the tops of the world's largest mountains. To put it into perspective, there are only about 4,000 species of mammals. (And almost 1,000 of those are bats; see page 222. Great news for anyone who loves Halloween.)

Spiders have hearty exoskeletons, too, and their ancestors do show up in the fossil record, often preserved in amber. Researchers concluded long ago that spiders descended from a many-legged, scorpion-like ancestor, scorpions being their next cousin in the overarching arachnid group. Genomics proved this relationship in 2017, and in 2018 researchers identified a piece of 100-million-year-old amber as containing a preserved proto-spider with, indeed, a long scorpion-like tail.

Outside of arachnids, the next relative is the insect group that includes the famous fruit fly *Drosophila menalogaster* (page 137), with whom spiders share similar jaw parts (mandibles). The next group over are insects, including the red flour beetle (page 65).

Spiders' path to their current-day level of diversity—that's where it gets a little fuzzy.

Around 700 million years ago, the group that would eventually include ticks and arachnids split off from the insect group that would come to include mosquitoes and the fruit fly.

Around 450 million years ago, the group that would become ticks split from the group that would become spiders and scorpions. We know this thanks to a massive international study that compared the genomes of ticks, spiders, and scorpions. They found what might be considered an evolutionary smoking gun: remnants of DNA that once belonged to a single ancient genome, the genome of a single ancient arachnid ancestor. This ancestor had reproduced via the unusual move of duplicating its entire genome, polyploidy, like the African clawed frog and the whiptail lizard. That whole-genome duplication, it seems, kickstarted the entire arachnid lineage.

For the social African velvet spider in particular, the spidery habit of hanging on to whole-genome information could be the reason for its persistence. This spider has a tendency to breed only within its own colony, so by all accounts, its populations should be inbred, which usually leads to genetic problems that can be devastating. Yet African velvet spiders thrive, and in a variety of habitats. In 2014, a group of researchers in China sequenced the velvet spider's genome, expecting it to be simple: inbred animals often have genomes that have pared themselves down to basics, having sloughed off overlapping information. Instead, the researchers found a long, convoluted genome with—indeed—a great deal of repeated DNA and transposons. Big, repeating genomes are a spider thing. Get used to it.

BEASTLY BREAKDOWN

MOUTH PARTS

The tools an animal uses to attack and catch and kill its prey play a huge role in its survival. Just as a researcher looks at the features of a vertebrate's jaw and teeth to identify it, arachnologists look at the shape and structure of a spider's mouth parts. Today, taxonomists classify species by their fangs, their sexual bits, other physical attributes, how they live, and how they hunt.

A group of 2,500 species has fangs that point straight down, including tarantulas, trapdoor spiders, and funnel-web spiders.

Another grouping of 97 species has horizontal mandibles (mouth parts), many of which also build trapdoors to capture prey.

As best as we can tell, the most diverse group is made up of jumping spiders (5,500 species), dwarf spiders (4,500), wolf spiders (2,400), and a few thousand web spinners. These categories arose from genetic analysis from the 1990s using the old shotgun method of study: pulling a few specific chunks of DNA at a time and comparing them. But those were shots in the dark.

VENOM

Spider venom is something humans have studied for a long time, because it helps us treat ourselves, should we get bitten. Each species of spider makes a different venom concoction of up to 1,000 different compounds.

Only a handful of venomous spiders can actually hurt humans, which is a wee fraction of the total number of spider species, but they get a lot of hype, and most of them live in Australia. The way venom works has even farther-reaching implications: the way venom targets channels of nerve communication can teach us a lot about nerve communication in ourselves, and insecticides that target specific insects (like the insects with which spiders have coevolved) can help humans hone in on certain pests without messing with whole ecosystems.

SENSORY HAIRS

As in all invertebrates, a spider's exoskeleton is made of a material called chitin. The fine "hairs" that cover its body are also made from chitin pushed out of tiny holes in its shell, though they are a little softer and more supple. Just as in a cat's whiskers, these hairs are hypersensitive and filled with nerves, and they serve as the velvet spider's primary sense.

EYES

Most species of spider, including velvet spiders, have eight eyes: four big ones and four little ones, arranged on their head to give them 360-degree vision and incredibly precise depth perception. (Some spiders, though, have zero eyes, and some have up to 12.) Each of those eyes is a camera-like eye, with a round pupil, just like ours.

Eyes of the *Epeira conica*, an orb spider with no common name. This chromolithograph by Vincent Brooks depicts the image magnified thirty times. For *Hardwicke's Science Gossip*, 1884.

As we'll see, though, eyes are a deeply seated animal trait, and they've evolved several different times in evolutionary history. Spiders and mammals definitely did not get our eyes from the same common ancestor, which would be weird.

But the truth might be weirder. According to genomic patterns, spiders probably started out with an early version of a compound eye, like those in other insects and aquatic invertebrates. Over time, these eyes separated from one another, distributed around the spider's head, and gave it excellent vision. Because they're unlikely candidates for excellent camera-eye vision, researchers have recently begun to dig into the evolutionary history of the spider eye to see how it might inform therapy for humans with genetic eye disorders like macular degeneration. In this case, if a spider could do it, maybe we could, too.

LEGS

Another hint that the spider isn't just another bug: its legs. It has eight legs, yes, and each of those legs has eight parts (if you count the claw at its tip).

Over time, the spider's legs became more specialized for jumping long distances. Vertebrates move our limbs like a catapult, using the muscles to push and pull the bones. A spider's legs work from the inside out, more like a hydraulic slingshot. Like all invertebrates, spiders' exoskeletons are filled with fluid. When it wants to jump, it pulls that fluid tight into its body, then releases it with a literal burst. The sudden burst shoots the spider's legs out and can get it from one place to another in the blink of a camera eye. This is why dead, dried-up spiders have their legs curled into their least resistant state.

BODY PLAN

Insects have three core body sections; ticks only have one. Spiders have two: the abdomen at its rear, and a combination head/thorax in the middle.

There's a string of genes that tells a spider's DNA how to align its body plan. If you look at this string of genes in an insect, they're organized linearly: first head, then hindquarters, then thorax with first set of legs, then second, third, and abdomen. Not so in spiders! Spiders develop the midsection first, with all eight legs and the head coming together at once.

THORAX

Fused to its head. Some colorful spiders have evolved sexually selected mating behaviors and use them to attract mates, à la Australia's peacock spider.

EGGS

Now that science has started paying attention to spiders, they've quickly become a favorite model organism for the study of evolutionary developmental biology. This is because their genomes look so different from other organisms, yet they still have some of the old-school research characteristics that model organisms share: they're small, they produce a lot of offspring and turn over generations quickly, and their eggs are translucent. This last feature is especially great for researchers in evolutionary developmental biology (or evo devo; see fruit fly, page 142). By spying on the goings-on of a developing egg, researchers can track changes that happen at each stage of development and compare them to one another across different animal species and even orders and families. The fact that developmental data on spiders can instantly join the fray bodes well for the future of spider research.

SILK

Genes code for proteins, which are clumps or strings of amino acids all connected together.

Spider silk is protein: a long string of amino acids stuck to amino acids. It's rare that we get to see the product of the thing that DNA codes for so clearly, spooling out of an animal's body. Another clear example is milk, which is exactly why researchers have previously tried to study the properties of spider silk by genetically modifying cows and goats to produce it from their udders.

Synthetic spider silk has long been considered a holy grail of engineering, even stronger than it is light and flexible. With the sequencing of the velvet spider's genome, researchers are able to read an exact recipe for the real thing, right there in patterns of As, Gs, Cs, and Ts (the initials of the four nucleic acids that make up DNA). But they're finding that even though it contains a great number of repeated sequences, it's incredibly long and even more complicated than they expected. The implication is that there are probably as many recipes for spider silk as there are species of spider. Let the untangling begin.

WEB SHAPE

In the years before genomics, web shape was one of the ways spider scientists would classify spiders. Arguments raged. Some scientists insisted that web shape was the key to understanding relatedness and evolutionary lineage, pointing to spiral orb webs (the classic shape, radiating out from a central point) and tangle webs (like cobwebs), both of which are most closely associated with two specific spider families, collectively considered orb spiders. But others would point to funnel webs, tube webs, and sheet webs, which are what they sound like and come from spiders across a variety of taxonomic groups. There are also spiders that don't make webs at all. In 2014, genome technology finally swung in to solve the mystery. A large American study comparing 40 different spider species found that orb spiders actually fell into two separate lineages. It wasn't web shape but type of silk that differentiated them: one group produces viscous silk, the other something called cribellate silk (which is more sticky and fuzzy, the kind that won't come off your sweater if you walk into it). The group of American researchers followed up with an even bigger study comparing 70 species of spider and found evidence of a big boom in spider diversification, much like the big bird boom (only much earlier). The group of spiders that makes viscous webs emerged during this time, as did most ground-dwelling spiders that don't make webs. The viscous-webbed orb spiders were more related to non–web makers than to web makers that made fuzzy webs, which possibly means that web making evolved *twice* in spiders. But more research is needed to follow the story exactly.

Interestingly, the big spider boom happened about 100 million years ago, around the same time as an explosion of nonflying insects. Another case of coevolution? Always. Now that's what I call a food web.

LESSON OF THE SPIDER:
We have a lot more to learn from the spider, evolutionarily and otherwise.

TERMS DEFINED: Whole genome duplication, evo devo

TEETH: great for bark-digging and grub-demolishing

EYES: great for seeing at night and looking terrifying

EARS: great for listening to grub-tapping

MIDDLE FINGER: great for grub-tapping, turns 360 degrees for grub digging

SCALE
3 feet
(91 centimeters)

FAM-O-METER

80%?

Aye-ayes are primates with about the same number of base pairs in their genome as we have—about 3 billion. But they seem to have stayed very differently arranged.

AYE-AYE

(Daubentonia madagascariensis)

PERHAPS OUR LEAST CUTE NEARISH RELATIVE

Aye-ayes are primates, which in a very general sense puts them in our same evolutionary family. But one look at their nightmarishly specialized features makes you think, *Well, that escalated quickly.*

WHEN IN (PRE)HISTORY/ WHERE ON THE RIVER

The aye-aye is one of nearly 100 species of lemur, all of whom live on Madagascar. Like the kiwi of New Zealand, it looks the way it does because it's on an island.

Today's primates include lemurs, tarsiers (who look like tiny, adorable lemurs), old and new world monkeys, gibbons, all the great apes, and us. About 55–65 million years ago, proto-primates split from the lineage that would become rodents. That common ancestor probably looked like a small rodent with good tree-climbing paws. From there, things started happening fast. Or so it seemed.

As Madagascar shifted farther from what used to be Gondwana, over the blink of a couple million years, lemurs diverged from other primates, and the aye-ayes diverged from most of the other lemurs.

The aye-aye is the most cosmopolitan of lemurs because it lives in three regions of Madagascar instead of just one. Each individual can occupy a range of 1,500 acres (600 hectares). This picture got even more refined in 2013, when researchers with Omaha's Henry Doorly Zoo and Aquarium compared the genomes from 12 aye-ayes from three regions and found that northern aye-ayes were genetically distinct from those in western and eastern regions.

BEASTLY BREAKDOWN

MAMMARIES

As mammals, aye-ayes have nipples. As primates, they have two. These ones, though, are situated more in the groin area.

EYES

Like so many nocturnal animals, aye-ayes have huge eyes that stare into your soul (or, rather, that maximize light intake). Maybe it wouldn't be so bad if they didn't also have those teeth.

TEETH

Aye-ayes have evolved to specialize at eating grubs out of dead trees—like a woodpecker, only terrifying. When it finds a potential meal, it digs into the wood with its teeth like the demon spawn of Nosferatu and a beaver. Like a beaver, its teeth never stop growing.

THAT FINGER, THOUGH (DIGITS)

As primates, aye-ayes have opposable thumbs and grasping hands. But one of these things is not like the others. Its middle finger is especially long and heavy, a built-in mallet. Even better, it then uses the same finger to reach into the wood it's chewed and dig out its prey: the finger bone turns around almost 360 degrees.

EARS

An aye-aye's big ears allow it to listen for grubs using a tapping technique. The researchers behind a 2017 genome study wanted to know if the aye-aye's genome had any overlap with genes associated with echolocating animals like bats and dolphins (pages 88, 222). It does not. The tapping technique isn't as developed a technique as full-on sonar, but it's a good example of

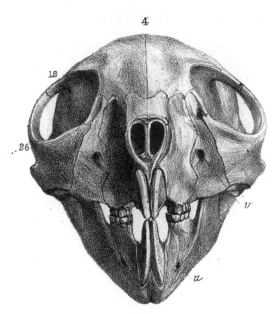

Skull, teet, and hyoid arch of the male aye-aye, 1863. Illustration by a physiologist and veterinarian with the actual name Johann Christian Polycarp Erxleben.

convergent evolution (see mini manatee, page 89) via different means. Recommendation for further study: look for genetic overlap with other grub hunters, like woodpeckers.

CHROMOSOMES AND GENETIC CHANGE

Despite these highly specialized adaptations, a 2012 genome study showed that the rate of change in the aye-aye's genome was fairly slow, actually, compared to other primates, mice, dogs, and a few other mammals. It also showed that aye-ayes have a relatively lower genetic diversity. This makes sense when you consider that they're genetically isolated; they only have half a set of chromosomes per parent, and there are only an island's worth of genes to choose from. Unlike *Xenopus* or neoaves, there

THROUGH DARWIN'S EYES

Once again, Madagascar was the territory of Darwin's colleague and conspirator A. R. Wallace. He said of the aye-aye in his 1876 book, *The Geographical Distribution of Animals:*

The Aye-aye, (Chiromys), *the sole representative of this family, is confined to the island of Madagascar. It was for a long time very imperfectly known, and was supposed to belong to the Rodentia; but it has now been ascertained to be an exceedingly specialized form of the Lemuroid type, and must be considered to be one of the most extraordinary of the Mammalia now inhabiting the globe.*

isn't much to change if there aren't as many options.

The story of the aye-aye reminds us that physical anomalies don't necessarily come from a revolution in genetic arrangement. Sometimes it's as simple as something works, and it keeps working. Despite its appearance, the aye-aye is, for now, just a good old-fashioned case of natural selection.

LESSON OF THE AYE-AYE:

Never forget that this weirdo is technically in our branch of the river of life. Also never forget how much it has changed since we all went our separate ways. The same is true for us, and for monkeys, and for apes. Some of us just turned out a little prettier than others.

Illustrated close-up of the aye-aye's highly specialized grub-getting finger. From the *American Journal of Science,* 1903.

WINGS: not so much

EGG MAKER: the size of a much larger bird's

FEATHERS: more like hairs

BEAK: more shrew-like than bird-like

SCALE
16 inches (40 centimeters) tall

FAM-O-METER

69%?

The human-kiwi comparison hasn't yet been run, so we could infer as follows: It's an archeosaur ancestor like crocodilians and birds, and leads a life far removed from most humans, unlike crows and chickens. Still, it has convergently evolved to have mammal-like traits. Perhaps the overlaps fall in the same range as its neighbor the platypus.

NORTHERN
BROWN KIWI

(Apteryx mantelli)

THE MAMMAL OF BIRDS

The bird whose name New Zealanders have taken to represent them has found its own little corner of its own little corner of the world. You might mistake it for a shrew, which is no coincidence...or rather, it's exactly a huge coincidence, otherwise known as convergent evolution (see page 39). New Zealand has no native mammals, so the ecological niche that might otherwise belong to a shrew was a perfect fit for this odd little bird.

WHEN IN (PRE)HISTORY/
WHERE ON THE RIVER

Ratites are a group of flightless birds that include Africa's ostrich, South America's rhea, elephant birds, Australia's emus, and—yes—New Zealand's kiwis. The kiwi's appearance is strikingly different from its distant cousins, which tells a story: The world's ratites emerged from a common animal ancestor, a population of proto-ratites. Scientists traditionally assumed these proto-ratites lived on the supercontinent Gondwana before it broke into South America, Africa, Australia, India, Antarctica, and the islands in between. But in 2014, a group of American researchers cross-referenced ratite anatomy with analysis of ratite mitochondrial DNA, and found that kiwis' closest relatives are the African ostrich and Madagascar's now-extinct elephant bird—but *not* the emu, from nearby Australia.

This more far-flung origin story gathered more wind beneath its wings in 2016, when a Japanese study of fossil and genetic evidence traced ratites' origins back to early flighted birds in North America, which was then part of the northern supercontinent Laurasia. In order to get all the way to the far side of Gondwana, some flying would have had to be involved. Though the chicken's predecessors may never have gotten off the ground, today's most famous flightless birds, it seems, did lose their wings from disuse.

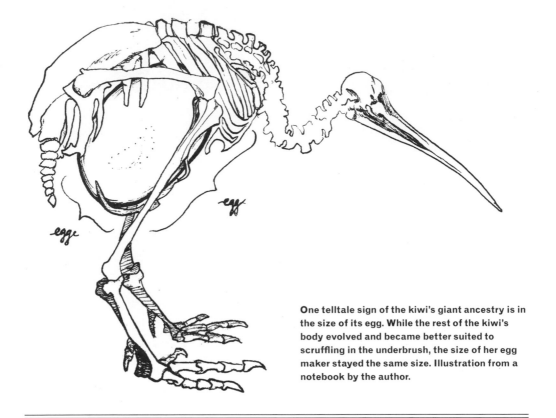

egg

egg

One telltale sign of the kiwi's giant ancestry is in the size of its egg. While the rest of the kiwi's body evolved and became better suited to scruffling in the underbrush, the size of her egg maker stayed the same size. Illustration from a notebook by the author.

BEASTLY BREAKDOWN

EGG MAKER

If you're skeptical that the relatively small kiwi comes from the same family as the taller ratites, just take a gander at the size of a kiwi's egg… right before it's laid. (Funny, right?) The absurdly exaggerated egg-to-body ratio shows that the gene sequences behind ovum and egg size function separately from the size regulation of the rest of the kiwi's body.

GIGANTISM

Most giants of the animal kingdom, such as the blue whale and the elephant, earned their size by finding a way to eat a lot of little things with low nutritional value. Ratites that remained large may have once existed solely on invertebrates—insects and their larvae.

EYES

Another thing kiwis lost over time was their ability to see in color. This loss seems to coincide with the kiwi's arrival in New Zealand, which means it's closely linked to its nocturnal, underbrush lifestyle. Only through more study will we know for sure if this loss (like its nocturnal lifestyle) preceded, followed, or is somehow genetically linked to its great sense of smell.

NOSE/BEAK

We don't usually think of birds as having noses; the tip of their beak is typically much farther south than their nostrils, which they don't much rely on for finding food. But the kiwi's nostrils have migrated down to the tip of its long beak, allowing it to maximize its sense of smell as it roots in the brush for insects and grubs. A recent study comparing the kiwi's genome to other archosaurs found that the kiwi has more genes associated with smelling than most, even its fellow nocturnal bird, the barn owl.

CIRCADIAN RHYTHM

The kiwi is the only ratite that is nocturnal, though we're not totally sure why. Whatever the catalyst, a kiwi's sensory systems have made the change accordingly.

FEATHERS

Completing its mammalian ensemble, the kiwi's feathers are long and thin, more reminiscent of fluffy hairs than feathers. They may also resemble Earth's earliest feathers, but only in appearance: early proto-birds had filamentous feathers, several loose filaments of feather emerging from a single point. Only later did some species evolve shafted feathers like we're used to seeing today, with a stiff central structure sprouting hundreds of fine barbs, so called because the filaments actually cling together via microscopic hooks. A kiwi's feathers aren't filamentous, although they might look it. Over millennia, their shafted feathers became shaggy and soft, hanging loosely like fur, which helps disguise it in the underbrush, just as it might any small mammal looking to live another day.

LESSON OF THE KIWI: A niche, like living in the underbrush and eating insects, is likely to be filled one way or another, such as underbrush being inhabited by small mammals. Animals that fill similar niches are likely to wind up sharing characteristics that behoove them in that niche, even if it winds up looking more like a creature from an entirely different order (like a bird that looks like a mammal).

TERMS DEFINED: Niche

For comparison: a common shrew (*Sorex araneus*). From the *Encyclopedia Brittanica*, 1866.

GALAPAGOS TORTOISE

(Chelonoidis nigra)

ENTER THE HOPEFUL MONSTER

The tortoise that inspired the original notion of a hopeful monster (an animal with some weird feature that just happened to work out). The Galapagos tortoise was also a primary inspiration for Darwin's theory, as you'll read below.

SHELL: an armored spine that grew way out of control

SEX ORGANS: can be determined by outside temperature

BEAK: not having teeth is a turtle thing from way back

MUSCLES: actually the slowest

WHEN IN (PRE)HISTORY/
WHERE ON THE RIVER

There are around 330 species of turtle and tortoise today. Even though turtles live both in and out of the water, turtles are traditionally classified as reptiles (their skin is dry instead of moist like an amphibian, and they hatch from leathery eggs, old-school, superficial stuff that mostly identifies what turtles/tortoises are *not* as opposed to what they *are*). More recent studies place them among the archosaurs: birds and reptiles like crocodilians.

360	250	MILLION YEARS AGO	65	0
PALAEOZOIC		MESOZOIC		CENOZOIC

...Diverged from archosaurs 255 million years ago. We know this thanks to a 2012 comparative analysis of the DNA and jawbones of four turtles, a caiman, a lizard, and a lungfish. A 2013 study even suggested that the turtle/tortoise lineage may have branched off before the other archosaurs, having as much in common with their shared predecessor as with either group (see hoatzin, page 106).

GALAPAGOS TORTOISE

...Arose from the tortoise lineage around 157 million years ago, probably walking to the islands from the mainland on a land bridge that hadn't yet submerged. As the Galapagos Islands themselves became isolated by the oceans, each island's tortoise population gradually evolved to look a little different than its long-lost cousins on other islands.

SCALE
Older adults can exceed 5 feet (1.5 meters) in length, 4 feet (1.2 meters) in height, and 550 pounds (227 kilograms) in weight.

FAM-O-METER

68%

Like crocodilians, turtles share origins with the chicken, and we all share a common ancestor from way back. But turtles have been doing a lot of specialized evolving through the millennia.

BEASTLY BREAKDOWN

SIZE

Among turtles, giant tortoises, like those made famous by Darwin (or that made Darwin famous), are the most monstrous. They used to live all over the world. Today, there are over 20 types, including those in the Galapagos and the Seychelles. Another species from islands in the Indian Ocean and still another Caribbean species just went extinct. The fact that all these types of tortoises live on islands tempts us to make some assumptions: these tortoises could only grow to such size in rich, protective ecosystems with few predators, like whales or elephants.

In fact, giant tortoises actually started out giant, tens of millions of years ago, when a lot more animals were giant. Prehistoric giant tortoises were even gianter, in fact, nearly twice as big in some cases. And they lived most places, including continental Europe, Asia, Africa, and the Americas. They even lived in colder climes and probably burrowed into the ground for warmth, much like their tiny turtle cousins, millennia later. As is the story with so many island-dwelling animals, the giant tortoises of today represent humbler versions of their ancient ancestors, quietly and slowly adapting to their specific environs.

LACK OF TEETH

The fact that turtles and tortoises have no teeth and have the facial appearance of elderly men is a complete coincidence and has nothing to do with science. But, in that vein, researchers have identified common patterns of gene loss.

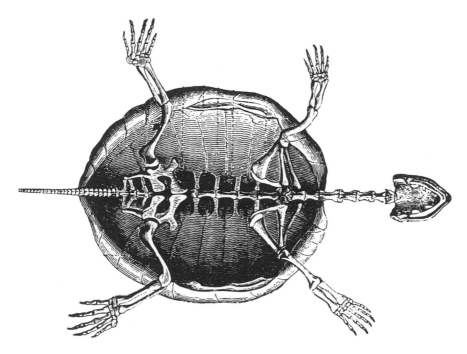

This 1887 illustration of the underside of a tortoise's skeleton illustrates how the specialized vertebrae fit together to form the inner layer of its shell. From Sarah Cooper's *Animal Life: In the Sea and on the Land: A Zoology for Young People*.

SEX DETERMINATION

Like some other reptiles, turtles can become either male or female in the egg, dependent on outside temperature.

SHELL

A turtle's shell is actually an extension of its spine. The monstrousness of turtles, namely their shell, has long defied taxonomists' efforts to peg their family history. So when did the shell come about? In 2018, researchers discovered a turtle fossil from 228 million years ago that shows a toothless turtle beak, a humorously long tail, and a round, frisbee-shaped body with the characteristic wide ribs and elongated vertebrae (holler, giraffe and bison) that would one day become a turtle's shell. But no shell. This is notable because some other prehistoric turtles have partial shells but no beaks. This is consistent with what we know about archosaur genomes: they're long, which must mean

Snakes, with their complete lack of limbs, can also fall into the category of hopeful monster.

something about their genomes keeps the disparate parts from interfering with one another, rather than jettisoning the information.

Think back to that wide variety of birds— the hoatzin for instance, which kept the claws and sad wings but gained more modern red feathers. This translates into the turtle's story like so: This archosaur had a lot of genetic information on hand, much of which might not go together or even seem useful—a large unwieldy ribcage and exaggerated spine versus a beak. But the turtle/tortoise made it work. Both ways. And eventually, the animals that kept both features won the day.

The soft-shell turtle study also found that its characteristically long, prehensile, almost shrew-like snout houses around 1,000 olfactory receptors. That's an unusually high number for a non-mammal. Just call them the kiwis of the deep.

SPEED

Tortoises have large, slow myosins—the opposite end of the spectrum from fruit flies and mice.

This also applies to their evolution. As a reptile, it shouldn't be surprising now that turtles and tortoises evolved slowly, about one-third as fast as humans. But even for reptiles, they bring up the rear: they evolved at one-fifth the speed of some pythons. The relationship between slow and long-lived makes sense if you think about it. The faster cells exchange energy, the more work they're doing, and the faster they're going to wear out. Like a car belonging to your teenage brother versus the car your grandma drives at 10 miles per hour to the grocery store twice a month. And the relationship between long life, long reproduction cycles, and slow evolution also makes sense, too. More time between generations means that it takes longer to pass along adaptations to their offspring. If generational turnover takes a long time, evolution will take a long time.

LONGEVITY

Throughout the animal kingdom, there's evidence to suggest that slow and steady does indeed win the race when it comes to life span. While some turtle species can live to 45 in the wild, other turtle species don't reach sexual maturity until their 40s and can live for well over a century. Some species of turtle can even become completely frozen during the winter, burying themselves and slowing down their functions.

HEART/BRAIN OXYGEN PROCESSING

Another thing that ages cells is the use of oxygen. Turtles, it seems, have adapted to survive on little oxygen, a thing that comes in handy when burrowing underground to escape the cold, just like their ancient ancestors. This information comes from the genome of the painted turtle, a globally ubiquitous turtle who's technically a tortoise. Its genome, sequenced in just 2013, was the first complete turtle genome, and only the second complete reptile genome, after the anole lizard.

The study identified 23 gene sequences related to the heart and 19 related to the brain that begin to express themselves when oxygen levels drop. The best part about these genes: humans have them, too. But unfortunately they don't activate when we're losing oxygen, as in the brain during a stroke or the heart during a heart attack.

Please note, however, eating turtles does not make you live longer or more likely to survive cancer, which is a misconception. And due to hunting and loss of habitat, half of the 330 species of turtle on Earth are currently endangered, making them the most endangered major group of vertebrates.

LESSON OF THE TORTOISE:

The tortoise is most famous for its shell, but it has a plethora of unseen adaptations that have kept it going strong for millennia and keep each tortoise going strong well into old age.

LITTLE BROWN BAT

(Myotis lucifugus)

FLYING UNDER THE RADAR

You might consider the little brown bat a hopeful monster in that it represents a revolution in evolution that worked out really well. But being a bat isn't a one-off like a turtle, an extreme endpoint. There are 300 species of turtle alive today—a slow, trickling evolutionary creek. In contrast, there are more than 1,100 known bat species, accounting for a full 25% of mammal species. The bat branch of the river is wide, and its waters run deep.

WINGS: basically huge webbed fingers

SCALE
Wingspan: 8.7–10.6 inches (22–27 centimeters)
Nose-to-butt: 2.4–3.9 inches (6–10 centimeters)

FACE: 'cause the eyes and nose lead food to the mouth

TEETH: rarely for vampiring

FAM-O-METER
70%

SEX ORGANS: can do some fancy acrobatics

WHEN IN (PRE)HISTORY/ WHERE ON THE RIVER

AROUND 85 MILLION YEARS AGO
...a third major split happened in the mammal world (third after monotremes and marsupials, then proto-elephants/manatees/armadillos/sloths). At this split, one lineage would eventually include mice and primates. The other would include proto-carnivores (cats and dogs), proto-odd-toed ungulates (horses and rhinos), proto-even-toed ungulates (giraffes, cows, whales, and dolphins)...and bats. All bats. This means bats share a more recent relative with all of those animals, including humans, than they do with mice, even though they look alike. This last major split sent bats on their own path. It was all bats from then on out.

AROUND 65 MILLION YEARS AGO
...like so many other animal groups, bats bounced back after the K-T event with a vengeance. Here, the diagram of bat family history starts to looks a lot like the explosion of birds in neoave.

ABOUT 63 MILLION YEARS AGO
...came the first intra-bat splits, around the same time that dogs split from cats. Scientists have classified bats in various ways through the years: echolocating versus not echolocating, bats with weird noses versus bats with weird rest-of-the-heads, bats who eat pollen versus bats that eat insects, cute bats versus ugly bats. Geneticists hoped that widespread genome comparisons would help clear things up. But by now we should know that genetics are more than skin-deep.

BEASTLY BREAKDOWN

FACE

The first split led to two major lineages whose fancy scientific names start with "Yin" and "Yang." On the Yin side, fruit bats split off first. Also known as flying foxes, fruit bats are large, about the size of house cats with wings to match. They have long faces, which once led scientists to assume they had evolved from the same lineage as lemurs. We now know that they are indeed related to all other bats, but just split off first and stuck with what worked. Today, even their cute faces have given them a little boost in survival. They look so adorable eating bananas in internet videos that humans will go out of their way to rescue and rehab them.

The rest of the Yin bats evolved to get smaller and eventually split into two more groups, one that would include bats with funny-shaped skin on their noses, like horseshoe bats and leaf-nosed bats, and one that includes bats with funny-shaped skin on their noses *and* that have tails that don't connect to their wings.

But meanwhile, on the other half of batdom (the Yang side), that lineage also split into groups that, superficially, were a free-for-all of free-tailed, weird-nosed bats that seemed to have traits show up or not show up with little to do with their relatedness. The conclusion: bat genes have retained a lot of options. It seems that a number of their traits emerge depending on need, as compared to true hopeful monsters that get stuck with a genetically fixed trait but do the best they can with it.

ECHOLOCATION

Biologists love to make a big deal of echolocation. If an animal even hints at being able to echolocate, someone organizes a study to find out how sophisticated their echolocation is and how detailed a picture it might be producing in the animal's brain. But like other superficial traits like wings or eyeballs or vocalization, echolocation comes about in many different ways and can emerge many times.

For instance, horseshoe bats and leaf-nosed bats both echolocate emitting sounds through their nostrils. Their nose skin shapes amplify the outgoing sound like a megaphone. But even flying foxes—with their pristine, seemingly acoustically impotent noses—echolocate, a 2014 study discovered.

Overeager geneticists did find something interesting, though, to do with the FOXP2 transcription factor, which is the one that helps control changes made in genes related to the brain/face/ear/vocalization connections. This contributes to speech in humans, songs in birds, and ultrasonic something-we're-not-totally-sure-about-yet in mice. In 2007, a team of Chinese and English researchers wanted to know if

A spear-nosed bat, found by Darwin in Brazil and illustrated by Captain FitzRoy of the *HMS Beagle*. Like many of Darwin's discoveries, the bat has a Latin name that has since been displaced by more specimens of more specific species.

FOXP2 had anything to do with echolocation. So they sequenced its analog in 13 species of both echolocating and non-echolocating bats, 22 other non-bat mammals, two birds, one reptile, and a platypus.

In echolocating bats, they found that arrangements of FOXP2 were extremely diverse—more diverse than across all the other animals except for humans known for having complex FOXP2 situations. Their level of complexity was on par with echolocating cetaceans like whales and dolphins. Some species even vocalize fake feedback sounds to confuse other bats into miscalculating and missing prey, leaving more food for them.

HAIR

Bats have hair and nurse their young, which makes them mammals. Their uniqueness among mammals means their genomes could inform all mammal genomes, including humans'. Their hair has made them key players in ecosystems as plant pollinators: they spread pollen on their hair just like bees and butterflies. Others eat fruit and spread the seeds in their guano (bat biologists' own special word for bat poop).

LONGEVITY

The little brown bat is an average bat, but they have lived to over 30 years old in the wild. This is an unusually long life for a small mammal. It's especially unusual when you consider its fast-paced lifestyle as compared with the turtle's slow movements, and its low metabolism.

IMMUNITY

Bats have an uncanny knack for carrying around pathogens that cause diseases that devastate other animal populations, while themselves remaining undiseased. Bats have been cited as the source of many an outbreak in humans and our animal familiars, including Nipah virus, Hendra virus, Ebola, severe acute respiratory syndrome (SARS), and rabies. But the complexity of their immune response–related genes keeps them safe.

Of all the body's systems, the immune system is the one best built to change and adapt. In fact, adaptation is its entire function. White blood cells collect information about intruding foreign bodies as they engage defenses against them. Then, not only does the immune system keep a record of that information in the body to ready itself to defend against future attacks, it actually makes and keeps little proteins particular to that exact invader. Those proteins are antibodies, and it's built into an animal's DNA to be ready to make antibody proteins of any shape, ready for any invader.

Well, almost any. Viruses, bacteria, and fungi are also on a mission to survive. As fast as animals evolve genetic systems against them, they evolve ways to trick those systems. Many viruses, for instance, evolve quickly enough that they become unrecognizable to our antibodies within just a few generations, which is why you need a flu vaccine every season. You've heard of antibiotic-resistant bacteria—that's due to bacterial evolution, too. (Not adaptation, mind: evolution. Change over generations.)

While the genetic intricacies are too nitty-gritty to get into, one recent story puts the bat's immunity story in real time. While bats have evolved to resist a wide variety of viruses and bacteria, one fungus has decimated them. Aptly named *Pseudogymnoascus destructans*, the fungus causes a disease called white-nose syndrome, which covers bats' faces and eats holes through their wings, like "athlete's foot on steroids," said one researcher. The fungus reduced certain populations of little brown bat by as much as 90% through the early 2000s. But by 2010, some little brown bats seemed to have devel-

THROUGH DARWIN'S EYES

By Darwin's time, the practice of dissecting animals and comparing their anatomy wasn't particularly novel. "Finding nearly the same bones as in bat's wings or man's hand" as he put it in *The Origin of Species*. Darwin understood that differing body parts shared a common origin in a common ancestor. He'd begun to understand that those common body parts diversified over generations via various selections. And that those body parts might end up serving shared purposes (flight, for instance) somewhat by coincidence.

oped (adapted or evolved to have) an immunity to the fungus's ill effects.

TEMPERATURE REGULATION

The little brown bat has the widest range of survival body temperatures of any known mammal—possibly of any known vertebrate. They can stay apparently unharmed at body temperatures as low as 43.43°F (6.35°C) and as high as 128.93°F (53.85°C).

REPRODUCTIVE CONTROL

Another uncanny characteristic of animals with long lives is their ability to control their reproduction. Female little brown bats not only can delay fertilization and implantation like some animals, but they can actually arrest embryonic development early in gestation. Here again is an animal with control over its genetic clock and long life.

The skeleton of a spear-nosed bat often used in academic texts, origin unknown.

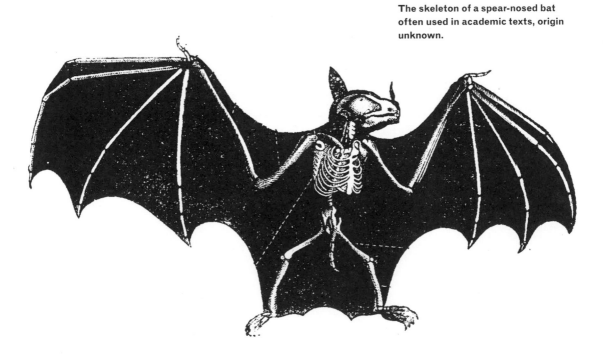

We don't know if all long-lived animals have reproductive control. We do know that not all animals that have reproductive control have long lives. Maybe all animals have a certain degree of reproductive control, but we just don't know it yet. (Certainly, we humans are not consciously in touch with that part of our biology.) But for now, it's either an active or inactive ingredient in the youth potion.

WINGS

Flight is the quintessential example of convergent evolution. A bat got its wings via a different path than birds and "bugs":

Birds and bats shared a vertebrate ancestor with non-flying limbs. Both evolved to turn their limbs into wings: one with the help of fancied-up body covering (feathers), and one with the help of extra skin.

Flying insects and bugs shared a much more distant eukaryotic ancestor with birds and bats that had shared body sections, which likewise (convergently) evolved into limbs and stubble, and one or both of those turned into wings.

(Please note: "flying" fish, "flying" lizards, "flying" squirrels and the like don't actually fly, they jump and glide, so we shan't count them here.)

The genetic changes that allowed for the change in hand shape are compound. Compared to the mouse's, the bat's genome shows unique changes in genes that determine which digits will be which and how many, genes that regulate digit length, and genes that make more or less cartilage (more in the bat's case). Especially curious is that the shorter digits are all tied to a gene called Fam5c, which sounds like a cell phone plan but is an important tumor suppressor. In other words, as much as bat genes grew the long finger, bat genes kept the other fingers short, using a gene usually employed to keep an animal from dying of cancer. That's a bat's MO: to take not dying to new heights.

LESSON OF THE BAT:
Adaptation happens, literally, on the fly.

PART 4:
THE SECRET TO ETERNAL LIFE

Living forever isn't the only measure of success. But it's a pretty darn good one.

Beating predators, getting fed, and finding a mate–that's kid stuff. In this section, we'll get to know some of nature's best-kept secrets to longevity from the inside out.

AXOLOTL

(Ambystoma mexicanum)

FOREVER YOUNG

Sure, it's cute, but this huge Mexican salamander can also take a lickin' and keep on kickin'.

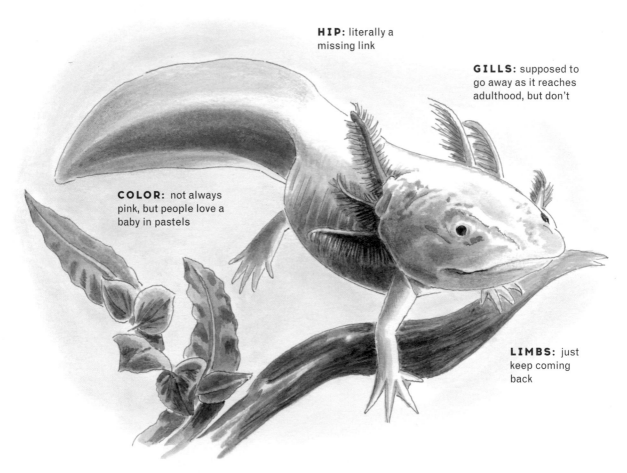

HIP: literally a missing link

GILLS: supposed to go away as it reaches adulthood, but don't

COLOR: not always pink, but people love a baby in pastels

LIMBS: just keep coming back

SCALE
16–18 inches (15–45 centimeters)

FAM-O-METER
76%?

For newer genomes, we really have to guess at overlap. The axolotl's a pretty distant relative of ours. But the sheer number of genes to choose from probably bumps up the percentage a bit.

WHEN IN (PRE)HISTORY/ WHERE ON THE RIVER

MESOZOIC	CENOZOIC

...About 90 million years ago, the amphibian lineage diverged ever so slightly from a common ancestor with lungfish.

AXOLOTL

...About 65 million years ago, the axolotl lineage split from the lineage of the giant salamander (over 4 feet [1.2 meters] long) and the lineage of the very bizarre Mexican mole lizard, aka the five-toed worm lizard, aka bipes, which looks like a worm with arms.

BEASTLY BREAKDOWN

COLOR

Most of the world's laboratory axolotls are descended from 34 animals that came to Paris from Mexico in the 1860s. They all have a form of undermelanization that makes them look pink. (This is reminiscent of the way that lab rats are usually white, except that lab rats are true albinos, as characterized by their pink eyes. True albino axolotls are yellow with darker pink eyes and gills.) Just as with cats, dogs, horses, chickens, guppies, and even humans, the range of colors in axolotl doesn't represent more than one species, just differences in a few gene sequences.

Now that humans are raising axolotls for pets and lab testing, the color combinations are likely to stray far from their natural mottle for generations to come. Its well-known pinkish hue sure lends itself to infantile comparisons. And maybe that's fair (foreshadowing alert).

PELVIS/HIP SOCKETS

If you recall from the section about lungfish, one of the most important innovations that led to the rise of tetrapods was a weight-bearing leg. And a weight-bearing leg requires a weight-bearing hip. Throughout scientific history, researchers have documented this transition using fossils, comparing embryological development, and eventually looking at DNA. But all of this work examined changes in the extremities, the part of the limb farthest from the animal's body. For a long time, science still needed to fill in the evolutionary gap between fish and tetrapod by filling the gap in knowledge between the foot and the hip bone. How did a fish's fin, however sturdy, reconnect with the spine enough to become weight-bearing? What might an evolutionarily transitional pelvis look like? A 2013 study of the axolotl found that it provided the perfect missing link. Turns out its weird state of arrested development, halfway between "tadpole" salamander and adult salamander, provided the perfect template against which to compare both fossil and genetic studies of hip development.

LIMB REGENERATION

All salamanders, some reptiles, octopuses, and a few fish have the power of regeneration. Even human infants can regrow the tips of fingers early in their lives.

In the late 18th century, a curious Italian observed regeneration over many species, including salamanders, tadpoles, snails, and earthworms. He even performed cruel if fascinating experiments on salamanders and found that all salamanders could regrow their tails. He also noted that they regrew faster in the summer than in the winter (although no current research can say for certain how temperature contributes to changes in cell information).

Axolotls, though, were special. They could grow back not just tails but all four limbs. And he had no idea how. His notebooks included drawings of teeny-tiny limbs inside the amputated limb waiting to grow.

Fast-forward to 2016, when an American researcher decided to devote her lab to understanding axolotl's miraculous power of regeneration. She found that after an amputation, the axolotl would bleed a surprisingly small amount and seal off the wound within hours. Soon, cells migrated to the wound site, forming a blob that was a little like a blood clot, only with skin cells instead of coagulating blood. The cells at this point were like embryotic stem cells: not yet sure what they needed to become. *Like* but not actually stem cells—this part was unclear. Was the axolotl pulling from a reserve of stem cells or changing existing cells to do its bidding? The first scenario is more likely, but where was it keeping its stash? Maybe the 18th-century Italian was onto something metaphorically with his mini-limb idea. However, once the cells got there, they quickly differentiated into bone, muscle, skin, and connective tissue, regrowing into a perfect limb, just like the missing one.

Most interestingly, when researchers cut out the bundle of not-exactly-stem-cells and placed it somewhere else in the body, it became a limb there, much like the fruit fly's wandering eyes or face-legs.

How or why did this useful but unlikely trait emerge in axolotls? Cannibalism! It's possible that for salamanders that start their lives in pools of hungry siblings, regeneration isn't just a cool trick, but necessary. That could be why they evolved the ability—or why they kept the ability while other animals lost it.

EYES

Axolotl regeneration extends to the lens of their eyes.

GILLS

The external gills share their origins and appearance with the axolotl's long-lost lungfish cousin. And they are a feature that most other amphibians have when they're younger, like a

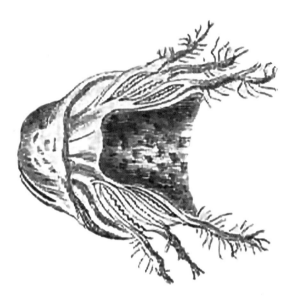

Close-up of gills, from field notebook of Thorson Gary, 1833.

frog in its tadpole phase. The strange thing about axolotls is that they keep their gills throughout their entire lives.

It's a risky statement that veers dangerously close to "why" territory, but it's tempting to make the connection between an axolotl's state of arrested development and its use of stem-cell-like cells to regenerate. All the more encouraging is the theory that axolotls evolved to hang out in their baby phase to offset the effects of fratricidal nibblings.

Here again, there's no literature I've yet found that will state the connections here outright, just as with the molecular effects of temperature. Which is prudent. It is not for science to say why. The more answers we get, the more hope we have of bringing limbs back to human amputees and maybe—just maybe—keeping that youthful glow way longer than is natural. If the genetic fountain of youth were easy to find, someone would have gotten rich off of it years ago.

LESSON OF THE AXOLOTL:
The key to regeneration might be in just never changing out of your birthday suit.

TERMS DEFINED: Regeneration

NAKED MOLE RAT

(Heterocephalus glaber)

ONLY AS OLD AS YOU FEEL

Sometimes, on Earth, the secret to success is getting naked, putting the needs of your friends in front of your own, and getting as snug as a bug that is smug because it's living longer than it should be for a mammal its size.

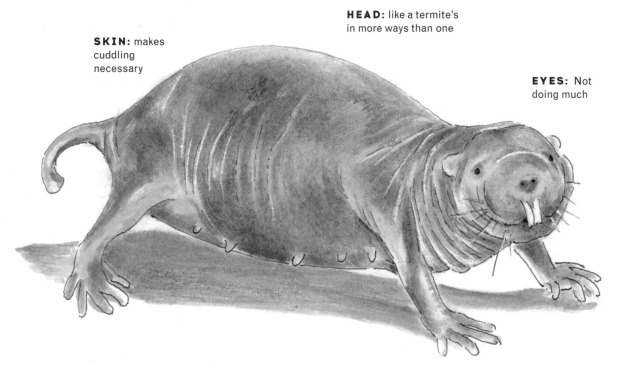

HEAD: like a termite's in more ways than one

SKIN: makes cuddling necessary

EYES: Not doing much

MOUTH: lips close behind big digging teeth for max impact in that tunnel-digging life

SCALE
About 2–3 inches (5–7.5 centimeters) long

FAM-O-METER

72%?

Somewhere between mice and bees.

WHEN IN (PRE)HISTORY/ WHERE ON THE RIVER

Native to East Africa, the naked mole rat is part of a larger group of pan-African mole rats, most of which have fur. Mole rats are a subgroup of the actually very large order Rodentia, which emerged about 79–89 million years ago. It includes everything from mice to capybaras, which can weigh up to 150 pounds. But then:

…About 24 million years ago, the naked mole rat started living underground. And the rest is history.

BEASTLY BREAKDOWN

MOUTH

Almost as horrific as the aye-aye's, the naked mole rat's teeth have evolved to an enlarged state. But instead of grubs, it uses them to eat tubers. And to dig—or gnaw if you prefer—through the dirt. Gross, maybe, but its lips close behind its teeth. And its teeth-digging is effective: naked mole rats can produce tunnels up to 6 miles long.

HEAD

Strong skulls and large heads help the naked mole rat find its way around among writhing hordes of its brethren, just like little rodent termites.

LONGEVITY

Naked mole rats were first discovered by a 19th-century French researcher who thought that they must all be very old. They were wrinkled and bald; why wouldn't they be? This superficial observation couldn't have been further from the truth. Naked mole rats can live into their 30s, which (some very small bats excepted) is unheard-of for a mammal that size. Relative to body size, naked mole rats live longer than elephants. So what's the secret to their longevity?

CANCER BLOCKERS

Most multicellular animals have some genes devoted to suppressing cancers. No naked mole rat has ever been found with cancer.

PAIN TOLERANCE

The phrase *no pain, no gain* sounds like some gym rat BS, but some research suggests the opposite might be true. Physical stress has been linked to the breakdown of tissue, stress that comes as sort of a feedback loop with pain. It's like the body is producing false pain to warn itself of potential damage, which the body then reacts to as damage. Naked mole rats, though, lack several genes associated with pain signaling. No pain, no stress, less damage.

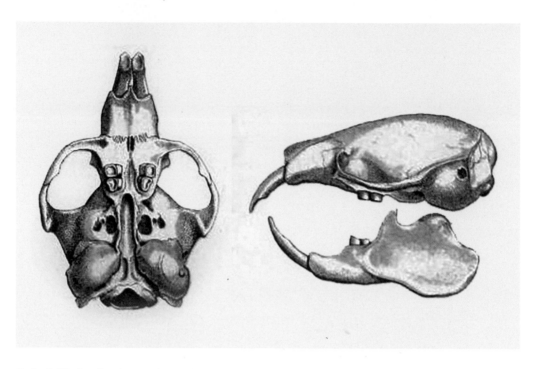

To the left is the view from underneath, minus the lower jaw of the naked mole rat; to the right are the upper and lower jaw in profile. *Proceedings of the General Meetings for Scientific Business of the Zoological Society of London.* 1885.

EYES

The Frenchman's discovery languished in a museum's archival drawer until, in the 20th century, an English researcher dug up his specimens and decided to look deeper into the animal. Examining living specimens, he found that the naked mole rat had turned off several genes related to vision since they live in the dark. He also saw a mutation in the gene dubbed "hairless," previously seen to cause baldness in mice and humans, which could explain how they lost their fur.

LOW METABOLISM

Naked mole rats don't do much, so they have low metabolisms, which also seems to contribute to longevity.

LOW-OXYGEN TOLERANCE

Living in sealed, moist, dark, dank, toxic, low-oxygen, high-CO_2 environments, naked mole rats have every reason to age before their time. Or do they?

Aging is related to the buildup of oxidizing agents in the body. Due to their subterranean lifestyle, naked mole rats can go 18 minutes without any oxygen at all.

EUSOCIALITY

Animal societies that have a distinct social hierarchy (think queen bee, worker bees) are eusocial societies. Eusociality is very common in insects, but the naked mole rat is the only eusocial mammal, which has afforded scientists a peek into the potential origins of it. Or, at the very least, its benefits. Eusocial animals, including insects, seem to be longer-lived than their closest cousins. Some pop-science research even suggests that the life span–related benefits of working together and having clearly defined roles in a community even extend to humans.

But the spontaneous eusociality in the naked mole rat seems to have emerged from it, at some point, becoming polyploid—having genes that come from multiple chromosomes. Which, in mammals, more often results in detrimental inbreeding. But then again, inbreeding in eusocial animals is common and actually may contribute to maintaining the social structure. Yet another evolutionary chicken and egg.

SKIN

That distinctive wrinkly, hairless skin was probably an adaptation that came out of inbreeding. Likewise, the naked mole rat can't regulate its temperature, like its fellow hairless animals, cold-blooded reptiles and amphibians. But thanks to its cuddly, underground, eusocial lifestyle, it's better off.

LESSON OF THE NAKED MOLE RAT: Who says insects should have all the fun/reap all the benefits of eusociality/claim this niche all to themselves?

TERMS DEFINED: Eusocial

DANCE MOVES: all in the mind

HAIR: made of exoskeleton

TONGUE: specialized for nectar-gettin'

POLLEN BASKET: a patch of prickly exoskeleton where the bee packs chunks of pollen to take home to her hive

FLOWER POWER: electrical charge generated by flight helps bees "talk" to the flowers

FAM-O-METER

61%

While a full comparison has yet to be made, parallels in genes related to social behavior in bees and humans may help us better understand autism.

HONEYBEE

(Apis mellifera)

HIVE MIND

BEASTLY BREAKDOWN

BRAIN

Bee brains have a million neurons in their head—one-thousandth the number in humans' brains. Honeybees live in societies that rival our own in size and complexity. A single hive may contain as many as 80,000 bees, which together build the hive, gather food, and feed the next generation of bees.

FLOWER POWER

Bees gather nectar from flowers, and they find flowers by merging many sources of information. This includes the position of the sun and the subtle nuances of a flower's scent and secret messages in ultraviolet signaling. Most mind-bogglingly, they share an electrical language with the flowers they pollinate/harvest from. Bees work up a strong positive charge while they're flying around. Flowers that have recently refilled themselves with nectar have a strong negative one. When a bee visits a flower, the two charges neutralize each other for a time. Coevolution at its finest (and most woo-woo).

EUSOCIALITY

In bee society, there are foragers, sterile female worker bees that tend to the larvae, male drones that mate with the queen, and the queen herself. These different kinds of honeybees might well seem like they belong to different species. The queen lives ten times longer than her workers, churning out 2,000 eggs a day. Yet the genetic information for creating all of these different types of bees is stored in the same genome. Each bee's fate is determined as it develops. All bee larvae are initially fed a substance called royal

A male drone, queen bee, and female worker bee. From *American Homes and Gardens magazine*, 1907.

jelly, secreted from the heads of the workers. It's a rich source of vitamins and other nutrients. It also influences how a bee develops. After three days, almost all the larvae get switched to a diet of honey. Only the queens in the making continue to get the royal jelly. Sequencing the honeybee genome could allow scientists to begin to piece together the way genes can help give rise to a complex animal society.

COMMUNICATION

Honeybees do something called a waggle dance, which researchers describe as the only known form of symbolic communication in invertebrates. When a worker bee finds flowers full of nectar, it returns to the hive and waggles out a "dance" whose pattern indicates specific details and directions as to the flowers' location. The other worker bees pick right up and copy the dance, then they take off in search of food, guided by their fresh, very precise, mental dance maps. All honeybees and only honeybees do this (as far as we know).

Over time, research has proven that making and reading the dance is totally instinctive for honeybees. Something in their brains automatically translates the route they've flown to return from the food source into a sort of dance code, and any honeybees that copy the dance automatically decode the route back. It's also true that all nine species of honeybee perform some form of the dance. But not all to the same degree. The dances of dwarf honeybees, for instance, are much simpler than the dances of the Western honeybees. All of this suggests the behavior is likely coded into their genome. The complexity of the dance indicates a specific lineage of adaptations. Some populations evolved in that direction, and others didn't. Dwarf honeybees, like crocodiles or lungfish, evolved early, and some of them just stuck with what was working.

Researchers still aren't sure exactly where the waggle dance lives in the honeybee genome. It probably has a lot to do with changes in the central complex (CX) of the insect brain, which is complicated enough as it is. Insect brains are

THROUGH DARWIN'S EYES

In a notebook referred to as "Notebook B," Darwin cryptically includes the bee in a thought about animals as interconnected and non-hierarchical.

All animals of same species are bound together just like buds of plants, which die at one time, though produced either sooner or later.—

Prove animals like plants.—trace gradation between associated and

nonassociated animals,—& the story will be complete.

And, on the next page:

It is absurd to talk of one animal being higher than another.—We consider those, when the intellectual faculties [|] cerebral structure most developed, as highest.—A bee doubtless would when the instincts were—

And here he trails off, and the record of the page ends.

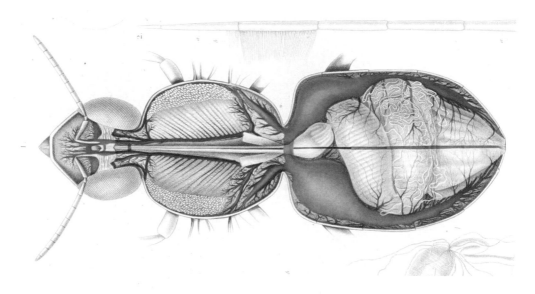

The internal anatomy of a honey bee. From *Le Règne Animal Distribué d'après Son Organization* (The Animal Kingdom Distributed According to Its Organization) by George Cuvier.

divided into two parts and connected in the middle by the CX, which is a bundle of different types of neural tissue, each with a special set of duties. For instance, some types of tissue are associated with compound eyes and possible light vision and place learning, and control motor impulses in legs and wings, while other bits likely have to do with that mysterious insect ability to find their own directional compass while flying through the sky. Together, the CX seems to be the place where understanding orientation and place gets translated into movement and recognition. In 2017, Australian researchers compared CX functions across all species of honeybees and found good reason to think that cracking the genome of the CX would help crack the code of the waggle dance.

HONEY DEW

Most people know that bees eat nectar. But they also eat at least one other macro-organism:

honey dew, which is a cute word for the juice that comes out of bug butts (*true* bug butts at least; see page 66). These are the types of bugs that live on and off of plants, sucking their juices and excreting what they don't need out the other end. In a 2018 study, researchers from the US isolated the percentage of DNA in the honey that was bug, bee, or flower. And found the percentage from honey-dewing bugs to be quite high. This has little to do with bees themselves, but I thought you should know.

LESSON OF THE BEE: Here's an animal that has evolved the ability to think conceptually, problem solve, communicate symbolically, and make food with its body. If you're not rethinking your definition of intelligence by now, you're more of a drone than these bees are.

AFRICAN ELEPHANT

(Loxodonta africana)

SKIN:
evolved to
stay cool

TRUNK:
prehensile
antenna
proboscis

SCALE

Males: 8.9–11.2 feet
(2.7–3.6 meters)

Females: 8.5–9.5 feet
(2.6–2.9 meters)

FAM-O-METER

72%?

While a deep comparison of the human
and elephant genomes hasn't yet been
done, similarity to dolphins and cows
might place our relatedness somewhere in between. What might really be
interesting is to compare the mammoth
genome with the genome of the early
human groups that hunted them.

WOOLLY MAMMOTH

(Mammuthus primigenius)

DEATH MATCH

These animals are close cousins, but one became famous for its longevity, while the other became famous for being extinct.

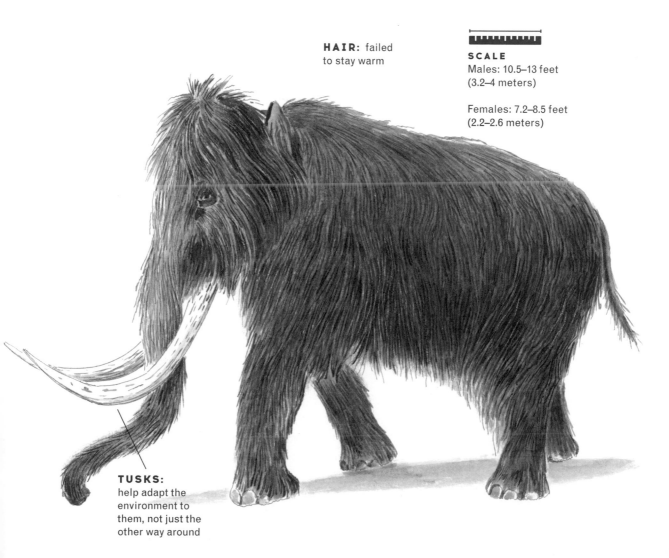

HAIR: failed to stay warm

SCALE
Males: 10.5–13 feet
(3.2–4 meters)

Females: 7.2–8.5 feet
(2.2–2.6 meters)

TUSKS: help adapt the environment to them, not just the other way around

WHEN IN (PRE)HISTORY/
WHERE ON THE RIVER

Around 90 million years ago, after another team sloth/armadillo/giant anteater stuck around in South America, team Africa again split into two. One group would eventually include dugongs, manatees, and the now-extinct stellar cow. The other would become mammoths and mastodons and would eventually give way to elephants. This latter group are called the proboscideans, for their long proboscises, or trunks.

BEASTLY BREAKDOWN

TUSKS

Before proto-elephants had trunks, they had tusks. Oddly shaped, specially modified teeth were another feature that was more common among prehistoric animals than in extant animals (although you'll still see them in walruses, wild hogs, and even a few species of toothed whales). Just like living elephants, mammoths probably used their trunks to dig for food and water; then getting at that water made longer and longer trunks an advantage. Elephants sometimes use their tusks to fight over mates, but perhaps most interesting is that they can use them to help them push over trees in places where mud holes or grazing areas would be more useful to them. This habit has made elephants one of the only other species besides humans that can greatly modify their environments.

BRAINS

We don't know much about the mental state of mammoths, but we know their brains were smaller than elephants', relative to body size.

Elephants have the largest brain of any living land mammal, and while brain size isn't everything, they possess a lot of those human-like markers we humans deem intelligent: play, learning, memory, and problem solving. According to an old-school animal psychology metric called the mirror test, they also understand that a mirror reflects themselves and not another elephant. For this test, researchers mark the test subject on the face with paint and put them in front of a mirror. Like great apes, and (possibly) dolphins, the elephants tested in this way examined the mark before trying to rub it off. In their case, with a trunk, of course.

TRUNK

Perhaps the most unusual thing about the elephant is its trunk: a long, highly sensitive, fully prehensile, mucus-filled tube that serves the same purposes as human hands and an insect's antennae. Elephants use their trunks to manipulate everything in their environment and may even use them to communicate over long distances by picking up low-frequency vibrations sent by stomping the ground. The trunk also houses a keen sense of smell, with more scent receptors than any other mammal.

SIZE

Hundreds of millions of years ago, being large was a lot more of a thing in the animal kingdom. Between 3 million and 100 million years ago, it was a lot more of a thing among mammals. Today, the only real giants left among mammals are great whales in the seas and elephants on land. Increasing evidence shows that humans might have played a major role in mammoths' extinction.

But before they survived humans, they had to survive the dice roll of their own genetics. The larger the animal, the more cells it has. The more cells it has, statistically, the more likely that some of them are going to become cancerous. But somewhere along the line, elephants evolved a series of genetic fail-safes to remind its cells to stop multiplying, effectively making elephants, against all odds, cancer-free.

BODY COVERING

Mammoths, on the other hand, did not win that genetic dice roll. A series of mammoth genome studies starting in 2015 found that woolly mammoth genomes just kind of fell apart over the course of several thousand years. In the same way that freak changes, deletions, and swaps wind up helping most animals better adapt to their environments, mammoth genomes started making mistakes that wound up causing it a litany of health problems, including short-circuited olfactory, digestive, and waste systems, and may have made certain groups sterile. Worst of all, some populations of mammoth inherited a defect that caused their hair to become fine, silky, and almost translucent, which just doesn't cut it in the thick of an ice age. And silky mammoth just doesn't have the same ring to it.

In contrast, the elephant grew into its new home in hotter climates. It lost its hair except for short hairs that mostly help it get feeling in its skin, which is 1 inch (2.5 centimeters) thick. They don't have moisture glands to help them stay cool but have instead evolved those characteristic wrinkles. When they pour water over themselves via their trunks, those micro-valleys help their skin retain the water 5 to 10 times as long as they might otherwise.

LESSON OF THE ELEPHANT:
This animal is in great shape inside and out, because neither it nor its genome forgets.

LESSON OF THE MAMMOTH:
This animal died cold and alone because neither it nor its genome could roll with the punches.

TARDIGRADE

(Hypsibius dujardini)

PERSISTENT LITTLE BUGGERS

Tardigrades seem to be the hardiest animal on Earth. We've long been fascinated by their ability to survive extreme habitats like subzero temperatures, boiling water, and even the heatless, airless void of space. We got so excited that we briefly let ourselves believe their abilities had otherworldly, preternatural causes that broke all the rules of life. And that may yet prove to be the case. But the more genetic science progresses, the more we understand that the rules of the genome aren't meant to be broken; they're just very, very complicated.

WHEN IN (PRE)HISTORY/ WHERE ON THE RIVER

Also known as water bears or moss piglets, fossilized tardigrades date back more than 500 million years. Science has identified 1,200 species thus far, living in just about any habitat you can think of—on land from desert to arctic, in freshwater, saltwater, and standing water alike.

Among the many species of tardigrade, there's a lot of genomic variation. The tardigrade pictured here, for instance, has a genome just half the size of another tardigrade species that's been heavily researched. This shouldn't seem *too* strange to you by now: looking at the African clawed frog, the whiptail lizard, and the velvet spider, we know that it's possible for animals to add or subtract a bunch of genome in the

evolutionary blink of an eye. Still, it's weird. And from what we know of the tardigrade genome, it is indeed weird.

The tardigrade has thus far ducked categorization by exhibiting a weird combination of the usual indicators of lineage. We know they're definitely invertebrates. This puts them in the same extremely general category as nematodes (roundworms like *C. elegans*) and arthropods (insects like fruit flies, spiders), but there's a big difference therein. Their body shape is more reminiscent of an arthropod, but a 2015 genomic study showed that they're missing a set of five HOX genes found in most other animals with HOX genes, including all arthropods. A

BODY: fully collapsible

EXOSKELETON: superpowered

SEX ORGANS: it all depends

MOUTH: tubular and telescoping

SCALE
0.002 to 0.05 inches (0.05–1.2 millimeters) long, but they usually don't get any bigger than 0.04 inches (1 millimeter)

FAM-O-METER

?

Until we do a side-by-side comparison with a large sample size for good measure, we just can't speculate on this one. We haven't even figured out what animals are tardigrades' closest relatives, much less how related they are to us.

HORIZONTAL GENE TRANSFER:
THE NEXT BIG FRIGGIN' DEAL

Horizontal gene transfer (HGT) is when DNA that started in one organism—even whole chunks of functional DNA (genes)—wind up becoming part of another organism's genome. We're not talking like cutting your hand and becoming blood brothers, or swapping spit, or whatever other crude exchange of bodily fluids. RNA and DNA are too smart for that; part of the job is knowing what belongs and what doesn't, just like antibodies in the immune system, only on a molecular scale instead of a cellular one.

HGT takes a few different forms, but the flashiest explanation is to picture a police procedural on TV. At the beginning of the show, the head of the forensics lab would report that there were two sets of DNA found at the scene of an armed robbery. The hilariously mismatched detective duo would have one suspect in custody and spend the whole show searching for the owner of the second set of DNA. But, a twist: In the final minutes of the show, you find out that the first suspect was in fact someone whose DNA had undergone HGT: *They had two sets of DNA in one body. There was never a second gunman.* Only, in real life the second gunman is usually just a virus or a bacteria. Most HGT occurs among microorganisms.

Organisms that have DNA from more

2017 study then found that their genes showed a lot of overlap with nematodes. But the same study showed that they showed telltale signs of rare genomic changes that would put them in the arthropod line. So…more research is needed.

This most recent finding comes after a whole lot of controversy over tardigrades' genetic makeup. Back in 2015, a major study concluded that 6,600 genes—one-sixth of its genome—was from other sources, including bacteria and even plants. The finding was cited in a respected peer-reviewed scientific journal and made headlines in some of the most respected popular science publications. But when a second research group sequenced the tardigrade genome, the number was less than 3%—within the normal margin of error. The 2017 study put that percentage at about 1%. The original finding was just bad methodology and contamination, which is a thing that happens. Especially with really old DNA (see Neanderthal).

Sometimes scientists get excited, though, which means sometimes they make mistakes. The reason everyone got so excited about the prospect of the tardigrade having DNA from other animals is understandable: *That's crazy, right?* Different species winding up with one another's

than one genome are rather informally called genetic chimeras, and they can become chimeras in a few different ways. Most of those complications occur right at the moment of change.

By now we've seen the many complications that can occur in an animal's DNA between the time it inherits it from its parents to the moment it passes it along to its progeny. In fact, the moment of inheritance is one such messy moment. Occasionally, a mother will retain some of her child's DNA after giving birth, making her a genetic chimera.

Another messy moment is illness or infection. DNA can be exchanged between parasite and host, or virus and host, or bacteria and their organic environment, or bacteria and bacteria and bacteria.

Or remember the demosponge and its ability to host a host of other organisms? And how that type of symbiosis happens everywhere across the kingdoms of life (animal, plant, microorganism, oh my)? Remember how sometimes those symbiotic relationships coevolve for so long that the organisms involved actually can't live or reproduce without each other? Like between flowers and bees or specific flowers and specific butterflies? Or between malaria and mosquitoes, then mosquitoes and their prey, then malaria and their new hosts? Or between humans and the good bacteria inside our own bodies? When organisms are living closely with one another, it's possible that somewhere, at some time, some of that DNA is going to get mixed up. (Especially when you consider how many ways RNA and DNA have to do their thing, as we've seen in this book.) This is probably what happened with the beautiful green sea slug, so newly discovered it's only known as *Elysia chlorotica.* Somewhere in its evolution, it absorbed DNA from the algae that probably made up most of its diet. Now, it photosynthesizes and gets some of its energy directly from the sun, like a plant.

Instances of animal-to-plant or animal-to-other-animal HGT are rare, so far as we know, but maybe that's only because we haven't known to look for it. Most of the HGT we know about occurs among microorganisms because their bodies are simpler and systems have fewer defenses; it's not much of a reach for a membrane to rupture, one to spill into the other, and organic chemicals to intermingle.

There's even a working theory that the first multicelled organism, whatever it was, became such when two single-celled organisms latched together. Picture an amoeba that eats a paramecium. Suddenly, the paramecium starts working for the amoeba, and both are more efficient. The amoeba's body becomes the cell wall of a new and complex cell, and then they replicate. Pretty soon, technically, you've got your first animal (see definition, page 12).

This is just a theory. But there is increasing evidence that HGT is yet another unusual method by which the genome has changed. HGT is proving to further confound the river of life.

actual DNA? That just doesn't happen. Right?

What's craziest is that this does happen. It's called horizontal gene transfer (HGT), and it doesn't happen a lot, but the fact that it sometimes does is fascinating. Animals are supposed to have all of the same DNA throughout their bodies. Pull a cell from anywhere and it reads like "this cell contains the DNA of animal X." HGT means that you might pull a cell from somewhere in animal X's body, and it reads as such, but if you pull a cell from somewhere else and it might read "this cell contains the DNA of animal X and also this rutabaga and also this slime mold and also this ancient archaebacterium." That's wild. But it happens.

Tardigrades are weird enough that HGT seemed like a reasonable explanation for some of the weirdness. As it is, we'll just have to accept that the weirdnesses came about by mechanisms we already have in our grasp. Or mechanisms yet to be discovered. Time (and diligence and research funding) will tell.

BEASTLY BREAKDOWN

EXOSKELETON

Tardigrades were discovered by a German pastor in 1773, and in 1776 an Italian clergyman and biologist discovered that water bears survive extreme conditions by making a transformation.

They can survive:

- temperatures as low as -328°F (-200°C) and as high as 304°F (151°C)
- freezing and/or thawing processes
- changes in salinity
- lack of oxygen
- lack of water
- levels of X-ray radiation 1,000 times the lethal human dose
- some noxious chemicals
- boiling alcohol
- low pressure of a vacuum
- high pressure (up to six times the pressure of the deepest part of the ocean)

MY ENTIRE EXISTENCE WEIGHS A TUN

When the going gets tough, tardigrades go into what is called a *tun* state: they collapse their bodies into a dried-up little package, yes, called a "tun." In a process much like deflating a bouncy castle, they go full carny and pack it in: suck all air and moisture out from between their tough tissues and fold in on themselves for an indefinite amount of time in cold (or hot or dry) storage.

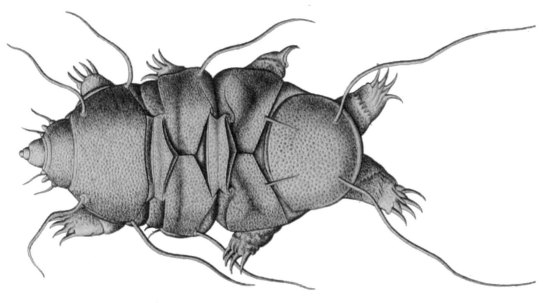

Echiniscus testudo species of
tardigrade, as drawn by Louis
Michel Français Doyère in 1840

HOW DO THEY DO IT?

COLLAPSIBLE BODY

Like some other hardy animals in this book
(turtles, frogs), one of the ways the tardigrade
survives is to collapse its body, tissue by tissue
and cell by cell, removing all air and water from
itself so as to have the most control over its
tissues' surfaces. When it's time to go under for
a while, the tardigrade retracts its weird little
mouth and limbs, and, expelling moisture out
its anus, collapses the rest of its body like an
inflatable pool toy that needs to fit back in its
box. In this state it can survive extreme
conditions better.

MOUTH

Tardigrades aren't picky about anything else, so
why would they be picky eaters? They can use
their gnarly telescoping tube of a mouth to suck

plankton out of water or suck the water out of
plants, or use their razor-sharp teeth to prey on
everything from nematodes to other tardigrades.

LEGS

A tardigrade has eight legs, arranged on four
sections, with a head on top. This coincidence
with arthropods is one reason researchers are
still inclined to want to categorize it there. But
that's some old-school platypus-identification
thinking; we know by now that eight legs could
be a product of convergent evolution, and the
difference in HOX genes between arthropods
and tardigrades throws the whole thing for a
loop. Each leg has between three and eight
spindly little claws, but whether the claws are
growing that way or breaking off remains to
be seen.

THROUGH DARWIN'S EYES

In 1872, 14 years after coauthoring their groundbreaking paper, Darwin wrote a charming letter to A. R. Wallace about *The Beginnings of Life*, a book by the controversial, problematic, and utterly fascinating biologist H. C. Bastian, famous for his propounding of "spontaneous generation," the possibility of living things spontaneously springing from nonliving things. In it, they discuss the tardigrade and rotifer, two animals so small and hearty that they seem to spring from nowhere. The reality, that the animals had been lying dormant for an astonishing amount of time, might have filled in some of the gaps between Darwin's frame of reference and Dr. Bastian's.

My Dear Wallace,—I have at last finished the gigantic job of reading Dr. Bastian's book, and have been deeply interested by it. You wished to hear my impression, but it is not worth sending.

He seems to me an extremely able man, as, indeed, I thought when I read his first essay. His general argument in favour of Archebiosis is wonderfully strong, though I cannot think much of some few of his arguments. The result is that I am bewildered and astonished by his statements, but am not convinced, though, on the whole, it seems to me probable that Archebiosis is true. I am not convinced, partly I think owing to the deductive cast of much of his reasoning; and I know not why,

but I never feel convinced by deduction, even in the case of H. Spencer's writings. If Dr. Bastian's book had been turned upside down, and he had begun with the various cases of Heterogenesis, and then gone on to organic, and afterwards to saline solutions, and had then given his general arguments, I should have been, I believe, much more influenced. I suspect, however, that my chief difficulty is the effect of old convictions being stereotyped on my brain. I must have more evidence that germs, or the minutest fragments of the lowest forms, are always killed by 212° of Fahr. Perhaps the mere reiteration of the statements given by Dr. Bastian [of] other men, whose judgment I respect, and who have worked long on the lower organisms, would suffice to convince me. Here is a fine confession of intellectual weakness; but what an inexplicable frame of mind is that of belief!

As for Rotifers and Tardigrades being spontaneously generated, my mind can no more digest such statements, whether true or false, than my stomach can digest a lump of lead. Dr. Bastian is always comparing Archebiosis, as well as growth, to crystallisation; but, on this view, a Rotifer or Tardigrade is adapted to its humble conditions of life by a happy accident, and this I cannot believe...He must have worked with very impure materials in some cases, as plenty of organisms appeared in a saline solution not containing an atom of nitrogen.

PROTECTIVE PROTEINS

When we're talking about animals this small, we've got to talk in terms of characteristics that are small as well, because that's where the magic happens. A tardigrade isn't hardy because it has special legs or skin; it's got survival mechanisms that work on a molecular level, protective proteins.

A protein found in at least one species of tardigrade, when inserted into human cells, can completely withstand dehydration and showed about 40% greater ability to suppress radiation damage. Researchers are hopeful that working with tardigrade proteins will lead to better methods of preserving organic material, like lifesaving vaccines that have to travel long distances or even lab-grown human tissues on their way to transplant patients.

REPRODUCTIVE BITS

Since tardigrades have taken the "whatever works" approach to staying alive, it's no surprise that they should take this approach to making new life as well. Some tardigrades reproduce sexually or asexually (via parthenogenesis, where a female lays and hatches viable eggs without males present; see whiptail lizards, page 100). The reasons for each mating approach probably have many complex factors: Some scientists say it varies according to species, but that could be because certain scientists observed certain mating habits in certain species—the same species might reproduce differently in a population or habitat, and under more or less cushy living circumstances (more or less food, water, hospitable temperatures). That would be a very tardigrade thing to do.

EGGS

Tardigrade eggs are pretty unsettling, with thin fibers that grow out like tentacles. Depending on the circumstances, a female might lay zero to three eggs at a time. Under certain circumstances, the female might lay them in the skin she just molted off, leaving them for a male to come along and fertilize soon after. Under certain circumstances, males and females seem to need to hurry up the act and engage in mutual stimulation: the female has to be stimulated to release her eggs, which she releases out under her outer layer of skin. The couple then stimulates the male, who releases his sperm into the same area. In this case, maybe the environment isn't a good place to leave her eggs, or maybe she wasn't ready to molt—most likely the exoskeleton provides nourishment or even additional genetic material for the developing eggs. Hatchlings come out looking like smaller adults—no pupa stage, just egg-to-tardigrade in—well, the length of time it takes depends on circumstances, too.

LESSON OF THE TARDIGRADE:

The tardigrade has some pretty far-out adaptations for surviving some pretty far-out situations. But even science can get a little too far-out when it comes to the idea of HGT. HGT happens. It's amazing. It just probably doesn't exist in the tardigrade.

TERMS DEFINED: Horizontal gene transfer, chimera

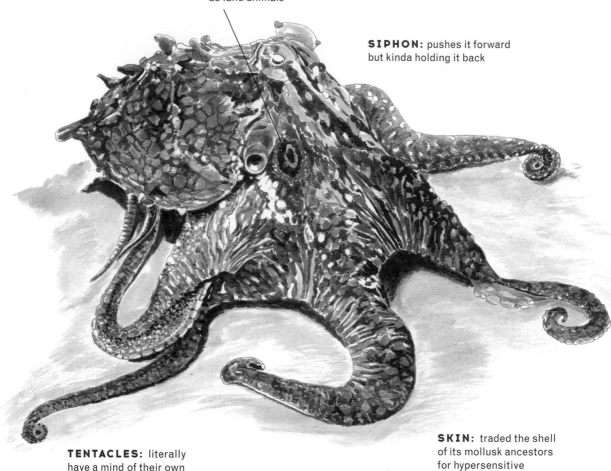

EYE: camera-like, same as land animals

SIPHON: pushes it forward but kinda holding it back

SKIN: traded the shell of its mollusk ancestors for hypersensitive super-camo

TENTACLES: literally have a mind of their own

SCALE
1 foot (0.3 meter) length

FAM-O-METER

24%?
38%?
43%?

Octopus edit their RNA within a single animal's lifetime, so the human-to-octopus comparison is an awfully complicated one to make. Like them, we might just be better off thinking up entirely new rules for an ever-expanding game.

CALIFORNIA TWO-SPOT OCTOPUS

(Octopus bimaculoides)

BROTHER FROM ANOTHER PLANET

Except it is actually still very much from this planet. If you haven't caught on by now, things can get plenty far-out right here at home. Right out of the gate, the octopus is breaking all the rules. We're looking at the California two-spot because in 2015 it was the subject of the first octopus genome project. The study's authors, a cross-institutional group of researchers from the US and Japan, chose the Cali two-spot because it was a relatively middle-of-the line octopus in terms of size, features, and habitat. It didn't seem to have any major outlying features compared to any other cephalopod. It's common throughout the entire Pacific, the biggest ocean in the world (not just around California). As the first cephalopod genome to be sequenced, it seemed like a simple place to start. The findings soon proved it to be anything but simple.

The 2015 study found that long stretches of this cephalopod's DNA and RNA had been changing throughout its life. Not just over millennia, not over many generations, not even from the last generation, but within this animal's lifetime. More insanely, within this animal's body. According to scientific understanding up until this point, one of the major tenets of the genome was that it was a blueprint for life: once it was set, it was set, the organism develops, is born, lives, grows, dies. There are changes during the lifetime of the organism, sure, but for the most part all the possible changes are spelled out in the organism's RNA and DNA beforehand, and at the most are switched on or off at different times, according to the organism's age, or maybe due to external factors. RNA editing, as the process is called, happens in other animals, but so far as we've known only to a limited extent. Humans have over 20,000 genes, but only a few places on about 20 of those genes are RNA-editable, and then only for the maintenance of small, physical parts. No RNA editing that's happening in our bodies within our lifetime would change our outward experience, and it would certainly have no guarantee of improving our lifestyle. Or so we've thought thus far. The notion that noticeable changes to our genomes might occur during our lifetime is a new notion known as epigenetics. The world's leading genetic experts still argue about the mechanisms of epigenetics, and some even doubting the validity of the notion at all.

But here comes an animal whose genome not only changes within its body according to its environment, it does it a *lot*. It's adaptation, even evolution, in fast-forward, the equivalent of the jump from Pony Express to email: evolution while you wait. Here again, the limitations of our vocabulary fail us. By Darwin's definition, this is too immediate—even seemingly too intentional—to be evolution. But he didn't have the technology to see as much of the picture as we do now. By the time he died, technology hadn't advanced enough for him to lay eyes upon any mechanisms for inheritance, not the gene, much less the stuff of which the mechanism is made (DNA) or the mechanism by which the mechanism gets made (RNA). Even so, he'd noticed patterns of artificial selection and natural selection. Were he alive today, what might he call this? Self-selection? Manual selection?

What kind of epigenetic changes is the Cali two-spot octopus making for itself, what kind of self-selection? Early follow-up studies show that it is editing to change proteins to accommodate temperature changes. A later genomic study of squids showed that they, too, make such edits, especially to build new nerves. "With these cephalopods, this is not the exception," said one of the authors of the squid study, "this is the rule. The rule is that most of the proteins are being edited."

All cephalopods branched out from a shelled cephalopod ancestor that was uncannily similar to the modern-day nautilus. And did you know cephalopods are mollusks? More specifically, they're part of a group called Lophotrochozoa, which also includes earthworms, leeches, snails, and shellfish like clams, mussels, and oysters.

The California two-spot octopus is one of around 300 species of octopus worldwide. They are most closely related to the common octopus, with which they've split off from the next branch over, a larger group of octopus including the webfoot octopus.

WHEN IN (PRE)HISTORY/
WHERE ON THE RIVER

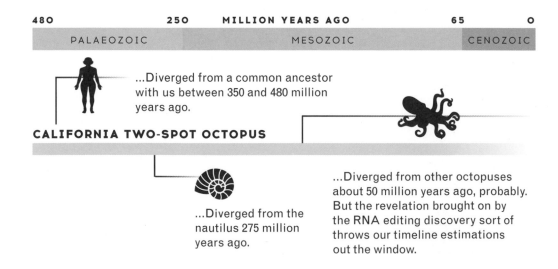

480   250   **MILLION YEARS AGO**   65   O

PALAEOZOIC   MESOZOIC   CENOZOIC

...Diverged from a common ancestor with us between 350 and 480 million years ago.

CALIFORNIA TWO-SPOT OCTOPUS

...Diverged from the nautilus 275 million years ago.

...Diverged from other octopuses about 50 million years ago, probably. But the revelation brought on by the RNA editing discovery sort of throws our timeline estimations out the window.

VAMPYRE SQUID

The larger branch of cephalopods on which Cali two-spot sits has a notable outlier with an enviably cool name: *Vampyroteuthis infernalis*. It literally means "the vampire squid from hell," so named because of its blood-red color and 360-degree interarm webbing that makes the whole animal look like it has devil wings. The octopus's misnaming was probably due to the two floppy fins at the top of its mantle, reminiscent of fins on a squid. (The mantle, by the way, is the big squishy round thing over an octopus' eyes that looks like its head but which actually holds all their organs and parts that aren't its tentacles.) But the vampyre "squid" share an ancestor with the rest of the octopuses, who had themselves already branched off from squid and cuttlefish.

The image here (from Ewald Rübsamen's 1910 book *The Cephalopods*) depicts the underside of the vampire squid, looking toward its beak, aka the last thing you'd ever see as a vampire squid's prey. Don't be fooled by the illustration: The tentacle "spikes" are actually specialized fingerlike suckers. The spike in the middle is the octopus's beak. Notably, the tips of the octopus's tentacles are bioluminescent (light up via chemical reaction), which helps it to attract prey.

THROUGH DARWIN'S EYES

Darwin was fascinated with octopuses, and though he lamented never seeing enough of them, he was able to glean some accurate insights from his few interactions with them. He had this to say about one of Cali two-spot's tropical cousins:

I was much interested, on several occasions, by watching the habits of an Octopus or cuttlefish. Although common in the pools of water left by the retiring tide, these animals were not easily caught. By means of their long arms and suckers, they could drag their bodies into very narrow crevices; and when thus fixed, it required great force to remove them. At other times they darted tail first, with the rapidity of an arrow, from one side of the pool to the other, at the same time discolouring the water with a dark chestnut-brown ink. These animals also escape detection by a very extraordinary, chameleon-like, power of changing their colour. They appear to vary the tints, according to the nature of the ground over which they pass: when in deep water, their general shade was brownish purple, but when placed on the land, or in shallow water, this dark tint changed into one of a yellowish green. The colour, examined more carefully, was a French gray, with numerous minute spots of bright yellow: the former of these varied in intensity; the latter entirely disappeared and appeared again by turns. These changes were effected in such a manner, that clouds, varying in tint between a hyacinth red and a chestnut brown, were continually passing over the body.

BEASTLY BREAKDOWN

EYE

Like all mammals, reptiles, birds, and fish, octopuses have camera-like eyes. But we vertebrates and cephalopods share a common ancestor from so long ago that eyes like ours weren't even yet in the mix yet. This means that the octopus and its cousins evolved to have eyes like ours independently from us, a strange phenomenon called convergent evolution.

This is similar to dolphin fins and shark fins: they wind up looking the same, but dolphins evolved from land mammals back into the sea, whereas sharks came from oceanbound fish like themselves. Our eyes are analogs with octopus eyes, not homologs, coming to the same end, serving the same purpose, but taking different paths to get here.

ABSENT SHELL

When early primates came to a fork in the road, proto-apes took the brawn path and proto-humans took the brains path. For cephalopods, one such fork was shell versus less shell. Some cephs with big, strong shells survived and passed on those genes, like the nautilus. Other cephs had smaller, weaker shells but were more nimble. On the less shell path, as each surviving generation lost shell, they gained speed. In modern-day squids, the thin, flat "shell" is entirely internal, serving only to give its long body some structure. Octopuses have mere vestigial shells at the back of their mantles. With all that speed and dexterity, cephs went from being the hunted to being the hunters.

INK

Just as Darwin observed, octopuses squirt ink when they're disturbed. The color clouds would-be predators' eyes, but toxins in it also irritate their eyes and muck up their sense of taste (which is like smell, underwater). Octopuses aren't immune, either: if they get stuck in a cloud of their own ink, they can die.

BLOOD

In almost all animals, the job of blood is to carry oxygen to all its body parts. In mammals, reptiles, birds, and even most insects, the blood cells use iron to help with that. Not the octopus. Its blood utilizes copper. Instead of iron-rich red blood, its blood is a deep teal blue (a deeper version of the green of copper rust, like the Statue of Liberty).

SIPHON

Octopuses evolved from mollusks, and mollusks have siphons, muscular valves that originally just helped them flush sand out of their soft innards. As mollusks evolved, they began using the siphons to move through sand or water via jet propulsion; those little holes you see in wet sand on a clam-rich beach are indications that a clam has flushed itself downward. By the time mollusks had fancy protruding legs and eyes, the siphon became sort of an outdated mode of transportation. But it's still the fastest way to get out of a jam, so they're still there, sticking out the side of an octopus's head, looking like a lopped-off tentacle.

HEART(S)

Octopuses have three hearts: two that work exclusively to move blood beyond the animal's gills, and a third that keeps circulation flowing for the organs. Ever since the octopus lost its shell, though, the traditional heart + veins + arteries system of internal hydraulics is sort of out of evolutionary sync with its incredibly squishable body. If all of this had been planned, it might do better with a free-for-all vein-free system like insects have, where all of their blood just kind of whooshes around inside their hollow exoskeleton. But as-is, the octopus has trouble keeping oxygen moving to all its parts when it has to undergo a major hydraulic shift in a short period of time, like when it uses its siphon to make a quick getaway. At times like that, its hearts, especially the one connected to the siphon, might temporarily stop. So it seems to prefer walking, hiding, or surprising its pursuer rather than siphoning away like its nautiloid relatives.

A close-up of the Cali two-spot's tentacle shows the fine-motor flexibility of each individual sucker.

MUSCLE TISSUE

The least-shelled of all mollusks, an octopus is so malleable it can slide its entire body through an area barely bigger than its own eye (in the Cali two-spot, that's about the size of a dime). Yet when that same "liquid" muscle rehardens, watch out. The average octopus is strong enough to lift more than its own weight, even under the full g-forces of land. One researcher describes watching an octopus collapse its body, creep up on a shrimp, extend one arm up and over the shrimp, and poke it, effectively scaring it right into its clashing beak.

An octopus's muscles are a totally different kind of muscle than our stringy ones (no need for bones to hold them in place or cartilage to muck up the works). It's called a muscular hydrostat, and each cell contributes both strength and shape, exchanging liquid instead of fiber to create motion. Humans do have one muscular hydrostat in their bodies: the tongue.

SKIN

The weird liquid movement of an octopus's muscles shows up in the skin, too. Depending on the species, it can turn smooth skin instantly rough when agitated, spiking into what look like spines or horns to scare off potential threats. As for color, Darwin was right: an octopus can make microscopic ovals on its skin shrink or flex to make the color and pattern it wants. Cross-institutional researchers found the skin of the cuttlefish *Sepia officinalis* (a close relative; see above) contains gene sequences "usually expressed only in the retina of the eye." Another type of cuttlefish can even mimic with its whole body the shape and movements of a hermit crab.

TENTACLES

All of an octopus's eight arms evolved from its ancestor's single foot. (Its fellow mollusks stuck with one: the slippery belly of a snail, the muscle-y meat of a clam, both are technically feet, too.) Somehow one became eight, although how is still a mystery. Each one of those eight arms is different from the others, or individuated, like your five fingers.

Each one is prehensile, meaning the octopus has motor control over it to the very tips, including each individual sucker. Tentacles even react after they've been severed. And if that happens, tentacles can be regrown.

"BRAIN"

Octopuses have the same type of neural wiring that makes up our central nervous systems, only for them, it's not centralized in a brain. Nerves that are more or less as complex as brain cells run through an octopus's body, a whole

three-fifths of it in the tentacles themselves. No wonder octopuses have the largest nervous systems among invertebrates: they basically *are* a nervous system.

All this talk about brain cells brings up the eternal question: are octopuses really intelligent? Intelligence is a complicated metric, not least of which because it's first and foremost a human metric. Humans are used to measuring intelligence against what *we* consider to be intelligent: *ourselves*. The more an animal acts like us, the more intelligent we consider it to be. Which, if you've read anything on this page thus far, should strike you as problematic when it comes to the octopus; at every turn, right down to their literal DNA, they're nothing at all like us.

Still, if we want to measure intelligence using some old-fashioned metrics, they'll play ball. Literally, if you'd like.

TOOL USE

The definition of tool use has come a long way since the scientific community standardized it in 1980 when it was a simple "using an object to solve a problem," geared to the notion of primates using sticks to dig for termites. Since that time we've seen crows disperse water volume using rocks and insects use each other as bait, so octopuses are a shoo-in for tool-user hall of fame. With brains in their arms, their real talent lies in manipulating objects, taking them apart, and escaping them, even after only being in touch with the object for a few minutes.

2008: At Sea Star Aquarium in Coburg, Germany, an octopus named Otto repeatedly extinguished the bright light shining into his aquarium by climbing onto the rim of his tank and squirting a jet of water at it using his siphon. The short-circuit had baffled staff and electricians until Otto was caught in the act.

2009: A California two-spot octopus at Santa Monica Pier Aquarium figured out how to loosen the valve on a spigot over her tank enough to flood the enclosure.

2009: Indonesian researchers document a California two-spot octopus carrying two half coconut shells around her body like a mobile home.

2014: At the Middlebury octopus lab in Vermont, a resident California two-spot octopus noticed a sea urchin was hanging around too close to its den, so she ventured out, found a piece of flat slate, and put it across the entrance to her den like a door.

2016: An octopus known as Inky slipped through a tiny opening in his enclosure at the National Aquarium in New Zealand and down a drainpipe that led to the ocean, and escaped to freedom (much like Hank, the fictional giant Pacific octopus in *Finding Dory*).

January 2018: A Maori octopus washed ashore inside a bottlenose dolphin, with a tentacle wrapped around its epiglottis (the thing that separates its windpipe from its food pipe) in just such a way that the dolphin had suffocated. Sure, they both died and the octopus can't have understood the exact implications of what it was doing, but it knew enough to keep hold of that body part and effectively killed its predator *from the inside*.

LESSON OF THE OCTOPUS:
Octopuses aren't from another planet but right here on Earth. They are masters of their domain.

WHICH REMINDS US, GETTING BACK TO THIS RNA EDITING THING...

It's not like the octopus is making these changes consciously. Right?

This is molecular-level stuff, a convenient mutation that basically speeds up other mutations, and there's no way the animal with those changes could have any conscious control over it. Right? Once again, this depends on what we mean by *consciousness*. And with the octopus, consciousness is a chicken-and-egg situation.

Which came first: Rapid-fire RNA editing that has certainly contributed to an animal's physical intelligence? Or a highly physically intelligent creature, adept at manipulating parts to solve problems. Could it be manipulating itself? No way. Right?

In the octopus, we see our brand of intelligence, but we also see an animal that is about as different from us as you can get. It freaks us out. (And now they can edit RNA? We only just got CRISPR up and running!) The differences between us are so vast that some researchers call the octopus "alien," which leads to misunderstandings on the internet. Octopuses aren't from another planet. They don't have to be. They've got everything they need to succeed right here on Earth.

AFTERWORD

The Russian writer Anton Chekov had a piece of advice he liked to dole out to his fellow playwrights.

In one version, he said, "One must never place a loaded rifle on the stage if it isn't going to go off."

At the beginning of this book, I hung the word *evolution* on the wall and told you it was loaded. And now I'm obligated to fire it.

All of this, everything you've read here—every animal, cell, and molecule—that's evolution. Loaded? It's *packed*. It's slow and mind-bogglingly complex. It's not just Darwin; it's the work of hundreds of thousands of researchers in an array of disciplines spanning centuries and counting. Every scientist and every study, every data point: that's just scratching the surface. Every unearthed fossil, every decoded genome, every cross-referenced gene sequence is one more piece to the puzzle, one more incredible thread in a fabric of all creation, if you'll allow the obscene mixing of metaphors.

We were talking about a gun. If the gun is evolution, I pull the trigger not to shoot anyone down but to blow people's minds. Ooo—okay no, like a warning shot, into the air, like a fun one to let everyone know it's safe to come back into the house and have a nice conversation over dinner. It'll be great. We're serving platypus.

WORKS CITED AND FURTHER READING

In regards to understanding evolution, science, nature, and the universe, this book barely scratches the surface. It scratches the surface of a body of knowledge (science in general) that is itself the very tiniest tip of an infinitely huge iceberg.

For instance: remember that additional category of life, the one that recently joined plants, animals, fungi, and bacteria on the river of life? In the few weeks since my editor blew her whistle and tore the manuscript from my claws, biologists may have found another one, thanks to sequencing the genome of a microscopic creature called a hemimastigote. Time and peer review will tell.

Point being: this work, and all works, should be considered in tandem with many others in the field. For recommended further reading for all experience levels and complete list of the 500 sources I referenced in writing this book, please visit *ctheplatypus.com*. And no matter what: Be skeptical. Stay curious.

SPECIAL THANKS

THANK YOU FOR EVERYTHING, ETERNALLY.

Raka Mitra
Barbara Morin
Dawn Frederick
Steve Freedberg
Justin Gregg
Steve Sweetman
Mary Voytek
Mary Roach
The #SciTwitter
    community at large
Susannah Schouweiler
Gina Svarovsky
Sarah Cohn
Julie Ann Pietrangelo
Dennis Cass
Eric Vrooman
Ashley Shelby
Doug Mack
Frank Bures
Sara Aase

Lars Ostrom
Sarah Schaack
Jessica Whithead
Jane Ryan
Charlie Sandford
Barbara Sandford
Fiddlehead Coffee
Caitlin Steitzer
Diego Martinez
Brittain Ashford
Alexandra Long
Sarah Moeding
Nic Vetter
Kelly Moritz
Anna Aslani
Chantal Pavageaux
Matthew Kessen
Jared Lodge
Jesse Ruuttila
Charlotte Hornsby

Amanda Svantesson-Degidio
Kate Morgan
Kate Sutton-Johnson
Kimberly Morales
Danny and Mariellen
    Sandford and the three
    weird sisters
Margaret and William
    O'Neill, McDowell
    and O'Neill clans
Logan O'Neill
Linda Brandt
Rachel DeCesario
Jan Freedman
Crystal Grimes
Peter Gaffney
Aldemaro Romero
Will Spottedbear

EXTRA SPECIAL THANKS: To Clark Sandford and Dinah Dunn, without whom this book would not exist.

INDEX

Illustrations are indicated in bold

ABS (Always Be Skeptical), 23, 31, 118, 120
Africa, 47–48
African clawed frog, 164, 165–167
  claws, 166
  DNA, 166
  evolution, 165
  eyes, 165
  larynx, 165
  senses, 166
  sex determination, 166
  sex organs, 166
African coelacanth, **157**, 158–161
  evolution, 161
  lobe fins, 161
African elephant, **242**, 244–245
  body covering, 245
  brains, 244
  evolution, 244
  size, 245
  trunk, 244
  tusks, 244
  vs. Woolly Mammoth, 243–245
African malaria mosquito, **110**, 111–113
  evolution, 111
  exoskeleton, 112
  extra senses, 112–113
  genome, 111
  proboscis, 112
  saliva, 113
  sex differences, 112
  wings, 112

allergies, 83
alligators, 197–199
al-Zahir Baybars, Sultan, 126
American crow, 90, 91, 92, 93
  brain, 91
  evolution, 91
  parenting, 92
  play, 92
  social learning, 93
  tool use, 92–93
  visual recall and counting, 91
American Plains bison, **94**, 96
  cow bits, 96
  DNA, 96
  evolution, 96
  genome, 95
  head, 96
  hump, 96
  interbreeding with cows, 96
  shared traits with cows, 95
  vs. cow, 94–95
animals, 12
  domesticated, 119
anteaters, 179
apes, 71
archosaurs, 92, 101, 108, 110, 197, 215, 217, 219
armadillos, 176–181
asexual reproduction, 103
Australian lungfish, 158
axolotl, **230**, 231, 232, **233**
  color, 231
  evolution, 231
  eyes, 232
  gills, 232–233

  limb regeneration, 232–233
  pelvis/hip sockets, 231–232
aye-aye, **208**, 209, **210**, 211
  ears, 211
  evolution, 209
  eyes, 210
  fingers, 211
  genetics, 211
  mammaries, 210
  teeth, 210

baleen whales, 34, 36–38, 41
barn owl, **200**, 201
  brain, 201
  ears, 201
  face shape, 201
  skull, 201
  talons, 201
  wings, 201
Bastian, H. C., 252
bats, 222–227
Beagle Laid Ashore, River Santa Cruz (illustration), **11**
bees, 238–241
beetles, 64–67
Beginnings of Life, The (Bastian), 252
birds, 106–109, 213–215
  evolution, 108
  genome, 109
bison, 94, 96
Blake, William, 143
blue whale, **34**, 35–37, **38**, **39**, 40–41
  baleen, 37

blowhole, 38
blubber, 41
evolution, 35
fins, 38
genomes, 36
hair/senses, 36
immune response, 38
low-oxygen tolerance, 38
pelvic area, 38
rostrum (snout), 37
size, 36
stomach, 39
tail flukes, 38
bonobo, **186**, 187, **188**, 189
  body shape, 189
  DNA, 187
  evolution, 187
  genome, 187
  immune response,
    188–189
  muscles, 189
  sexual proclivities, 189
  social intelligence, 189
bottlenose dolphin, **84**, 85–87
  body shape, 86
  brain, 86
  communication, 86–87
  evolution, 85
  rostrum (snout), 86
  sonar, 86
  tool use, 87
bovines, 96
butterflies, 57, 59

C. elegans, 62, **63**
California two-spot octopus,
  254, 255–256, 258–259,
  260, 261–263
  absent shell, 259
  blood, 259
  "brain," 261–262
  DNA, 255–256
  evolution, 256
  eye, 258

genome, 255–256
heart(s), 259
ink, 259
muscle tissue, 260–261
RNA, 262
siphon, 259
skin, 261
tentacles, 261
tool use, 262–263
cancer, 74–74
cat, **124**, 125–126, **127**, 128,
  129, **130**, 131, **132**, **133**
  brain, 128–129
  claws, 131
  coloring, 130–131
  DNA, 125–128
  ears, 128
  evolution, 125–127
  eyes, 129
  genome, 128, 130–131
  jawbone, 128
  kidneys, 130
  metabolism, 133
  nose, 128
  senses, 131
  sex organs, 131
  whiskers, 128
cephalopods, 254–263
cetaceans, 40
Chekov, Anton, 264
chickens, 192–195
chimpanzees, 187–189
chromosomes, 102
clones, 102–102, 103, 105
coelacanth, 157–161
coevolution, 59, 60, 150
comb jelly, **152**, 153–154, **155**
  cilia, 154
  colloblasts, 154
  neurons, 154
comparative anatomy, 40
convergent evolution, 39, 89
cows, **95**, 96–97, **98**, 99
  DNA, 96

evolution, 97
genome, 95
height, 97
immune response, 99
interbreeding with
  bison, 96
metabolism, 97
milk, 98
muscle, 98
shared traits with
  bison, 95
vs. bison, 94–95
CRISPR gene editing system,
  112, 262
crocodiles, 196–199
crows, 90–93
  genomes, 93
culture, 87
Cuvier, Georges, 178

Daphnia, **62**, 63
Darwin, Charles, 11, **13**, **14**,
  15, 16, 19, 22, 27, 33, 34,
  40, 42, 45, 51, **54**, 59, 61,
  71, 75, 77–78, 87, 103, 113,
  119, 120, 123, 133, 138,
  155, 167, 168, 175, 178,
  195, 211, 216, 218, 226,
  240, 245, 252, 256, 258,
  261, 264
  caricatures of, **15**
  on horses, 54
  "Notebook B," 240
  theories, 14–17
  trees of life theory, 16,
    17, 34
Delphinus FitzRoy, 87
demosponges, 146–151
Denisovan, 80, 81
  DNA, 83
  genome, 83
Descent of Man, The (Darwin),
  133
diabetes, 79

dinosaurs, 27, 36, 92, 101,
    106, 108, 169, 173, 193,
    194, 196, 197
Diogo, Rui, 189
dire wolf, **116**
DNA, 44, 48, 53, 58, 62, 72,
    79–81, 83, 102, 119,
    125–128, 166, 248–250,
    255–256
    manipulation, 75, 112
    mutation, 82
dogs, 114–123
dolphins, 39, 40, 73, 84–87
    genes, 85
domestic dog, **114**, 115–119,
    120, 121–123
    brain, 119–120
    circadian rhythm, 123
    coat, 120
    DNA, 117
    ears, 121
    evolution, 115
    eyes, 122–123
    feet, 122
    humans and, 116–117
    jaw and teeth, 121
    nose, 121
    rolling in stink, 121
domestication syndrome, 123
duck-billed platypus, **20**,
    21–23, **24**, 25, **26**, **27**
    bill, 23
    eggs, 25
    eyes, 23–24
    feet, 24
    genomes, 22–25
    hair, 22–23
    lineage, 21
    mammaries, 26–27
    plural of, 21
    skeleton, **27**

Egypt, 126
eukaryotes, 72

evolution, 10–11, 264
    speed of, 38
    study of, 12, 38

Feejee Mermaid, 22, **23**
Fertile Crescent, 126
finches, 45, 168–171
FitzRoy, Captain, 87
Floros, Joanna, 184
fowl, 192–195
FOX gene, 75, 171, 185
FOX2p factor, 224–225
foxes, 115
frogs, 164–167
fruit flies, 111, **136**, 137–138,
    139, **140**, **141**, 142–143
    alcohol tolerance, 143
    brain, 143
    cheapdate, 143
    DNA, 137
    evolution, 137
    eyes, 138–139
    genome, 137
    heart, 142
    myosin, 142
    stubble, 142
    thorax, 139–140, 142

Gadow, Hans, 167
Galapagos Islands, 45, 168,
    216–221
Galapagos tortoise, **216**, 217,
    218, **219**, **220**, 221
    evolution, 217
    genome, 221
    heart/brain oxygen
        processing, 221
    lack of teeth, 218
    longevity, 221
    sex determination, 219
    shell, 219
    size, 218
    speed, 221
gender vs. sex, 105

genes, 50, 58–59, 71–72, 75,
    119, 248–249, 264
genetics, 44
*Geographical Distribution of
    Animals, The* (Wallace), 211
Gharial, **196**, **197**, **198**, 199
    evolution, 197–198
    eyes, 199
    feet, 199
    ghara, 198–199
    gizzard, 199
    legs, 199
    longevity, 199
    rostrum, 199
    scales, 199
    sex determination, 199
giant marine isopod, **68**, **69**
    eggs, 69
    eyes, 69
    sex, 69
    size, 69
giraffes, 46–51. *See also* Masai
    Giraffe.
Goldschmidt, Richard, 191
Gondwana, 209, 213
Gray short-tailed opossum,
    172, 173, **174**, 175
    evolution, 173
    eyes, 173
    immune response, 174
    milk, 173–174
    saying name, 174
    sex organs, 174
    tail, 175
    transposons, 175
    whiskers, 175
Great Barrier Reef demosponge,
    146, 147–151, 154
    body plan, 149
    circadian rhythm, 151
    collagen, 150
    evolution, 148
    digestion, 149
    immunity, 150

neuron genes, 151
proto-tumor suppressors, 151
reproduction, 149
skeleton, 149
symbiosis, 150
tissue, 149
grizzly bears, 29, **31**, 32, 33

H.M.S. Beagle, **11**, 13, 14, 22, 113, 178, 245
Henry Doorly Zoo and Aquarium, Omaha, 209
hermaphroditic reproduction, 102
Hoatzin, **106**, 107–109
  evolution, 108
  feathers, 107
  foreclaws, 107
  metabolism, 107
  sounds, 107
  stomachs, 107
Hoffman's two-toed sloth, **177**, 178, **179**, 180–181
  circadian rhythms, 181
  claws, 180
  defenses, 180
  evolution, 179
honey bee, **238**, **239**, 240, **241**
  brain, 239
  communication, 249–241
  eusociality, 239–240
  flower powers, 239
  honey dew, 241
"Hopeful" monsters, 191
horizontal gene transfer (HGT), 248–249
horse, **52**, 53, **54**, **55**
  evolution, 53
  hooves, 55
  legs, 55
  teeth, 55
house mouse, **182**, 183–184, **185**

blood, 184
DNA, 184
ears, 185
evolution, 183
gene interaction, 184
muscles and myosin, 184
sex differences, 184
sex organs, 184
HOX genes, 140, 142
humans, **70**, 71–72, **73**, 74–75, 116–117
  alcohol tolerance, 143
  brain, 73–74
  communication, 75
  DNA, 72, 83
  dogs and, 116–117
  evolution, 71
  genome, 71–72, 75, 80, 83, 135
  heart plaques, 74
  HSL genes, 73–74
  male genitals, 73
  medical intervention, 75
  population shape, 73
  reproduction, 73
  salivary gland, 73
  sweat, 73
  wolves and, 118
hybrids, 30

immune cell receptors, 83
Improbable Destinies (Losos), 10
insects and plants coevolution, 59
invertebrates, 72
isopods, 68–69

Japan, 65
jawbones, 160
jellyfish, 153
jungle fowl, aka domestic chicken, 192, 193, **194**, **195**
  beak, 194

evolution, 193
forelimb, 194
genome, 195

kiwi, 212–215
K-Pg event, 108, 169, 223

lemurs, 209–211
Linnaeus, Carl, 30, 36,
little brown bat, **222**, 223, **224**, 225, **226**, 227
  echolocation, 224–225
  evolution, 223
  face, 223–224
  hair, 225
  immunity, 225
  longevity, 225
  reproductive control, 226
  temperature regulation, 226
  wings, 227
lizards, 100–105
London, England, 58
LUCA (last universal common ancestor), 72
lungfish, 156, 158–160

macromutations, 191
Madagascar, 209–211
malaria, 111, 113
mammals, 20–27, 28–33, 34–41
manatee, **88**, 89
  evolution, 89
marsupials, 172–175
Martens, Conrad, 11
Masai giraffe, **46**, **47**, 48–51
  blood vessels, 50
  coloring, 48
  diaphragm, 50
  evolution, 47
  heart, 50
  neck, 48–49
  skeleton, 48–50

Masai giraffe *(cont.)*
   stomach, 50
   vocal cords, 50
Mayr, Ernst, 78, 79
Mechanism of Mendelian
  Heredity, The (Morgan), 138
MEGA (software), 137
Megatherium americanum, **178**
Melville, Herman, **41**
Mendel, Gregor, 45, 138–139
Moby Dick (Melville), 41
model organism, 63
mole rats, 234–237
Morgan, Thomas Hunt,
  138–139, 142
mosquitos, 110–113
   evolution, 111
moths, 57–61
Muhammad, 126
mutation, 82

naked mole rat, **234**, 235,
  236, 237
   cancer blockers, 235
   eusociality, 237
   evolution, 235
   eyes, 236
   head, 235
   longevity, 235
   low metabolism, 236
   low-oxygen tolerance, 236
   mouth, 235
   pain tolerance, 235
   skin, 237
National Aquarium, New
  Zealand, 263
natural selection, 19, 40, 61,
  75
Nature, 16
Neanderthal, **76**, 77, **78**,
  79–81, **82**, 83
   body shape, 79
   coloring, 79
   diabetes, 79

DNA, 78–81, 83
   evolution, 78–79
   genome, 83
   head, 81
   sleep patterns, 81
   tobacco addiction, 81
Neaves whiptail lizard, **100**,
  101–105
   evolution, 101
   genome, 100–101, 105
nematode, 62, **63**
neurons, 151, 154
New Zealand, 213–215
nine-banded armadillo, **176**,
  177, 178, 179, **180**, 181
   circadian rhythms, 181
   claws, 180
   defenses, 180
   evolution, 179
   vs. Hoffman's two-toed
    sloth, 176–181
Nobel Prize, 138, 142, 143
northern brown kiwi, **212**,
  213, **214**, 215
   circadian rhythm, 215
   eggs, 214
   evolution, 213
   eyes, 214
   feathers, 215
   gigantism, 214
   nose/beak, 215
Nüsslein-Volhard, Christiane,
  140, 142

observation, 19
octopus, 254–263
On the Expression of Emotions
  in Man and Animals
  (Darwin), 120, 133
On the Origin of Species
  (Darwin), 15, 19, 33, 40, 51,
  71, 75, 77, 119, 226
opossum, 172–175
organism, 72

Owen, Richard, 27
owls, 200–201

Pangea, 167
Pax6 gene, 139
peacock spiders, 205
peppered moth, **56**, 57–61
   chemical senses, 61
   coloration, 58
   evolution, 57
   eyes, 60–61
   metamorphosis, 61
   poison control, 59
   proboscis, 59
   respiration, 61
   play, 86
polar bears, **28**, 29, **30**, **31**,
  32, 33
   evolution, 31
   feet, 33
   genomes, 29
   hair, 33
   head, 32
   metabolism, 32
   neck, 33
   size, 32–33
   tail, 33
  porpoises, 87
  positive selection, 49
Preservaton of Favoured Races in
  the Struggle for Life (Darwin),
  19, 71

raptors, 201
red flour beetle, **64**, 65, **66**, 67
   behavior, 67
   cannibalism, 67
   evolution, 65
   metamorphosis, 66
   mouth parts, 66
   sheathed wings, 66
regeneration, 232–233
reptiles, 100–101, 199, 217,
  219, 221

RNA, 44, 72, 82, 112, 184, 255–256, 262
rodents, 234–237
rotifer, 252

Saint-Hilaire, Étienne Geoffroy, 25, 26
Santa Monica Pier Aquarium, 263
scientists, 22
Sea Star Aquarium, Coburg, Germany, 263
sex determination, 102
sex vs. gender, 105
Shaw, George, 23
shrimp, 144–145
simians, 186–189
SIV (simian immunodeficiency virus), 188–189
sloths, 177-181
snakes, 219
South American lungfish, **156**, 158, **159**, 160
    evolution, 158
    gills, 159
    immunity, 150
    lobe fins, 159–160
    lung, 159
    vs. African Coelacanth, 156–161
species, 19, 30, 78
Spencer, H., 252
spiders, 202–207
sponges, 146–151
spontaneous generation, 252
symbiosis, 150

Tamura, Koichiro, 137
tardigrade, 246, **247**, 248–250, **251**, 252, 253
    collapsible body, 251
    DNA, 248, 250
    eggs, 253
    evolution, 246

exoskeleton, 250
genome, 246
legs, 251
mouth, 251
protective proteins, 153
reproductive bits, 253
teleosts, 43
telomeres, 171
theories, 14
Torres, Manuel, 18
tortoises, 216–221
traits, 145
transposons, 58, 61, 62, 111, 184
trees of life, 16, **17**, 34
Trinidadian guppy, **42**, 43–45
    coloring, 43
    evolution, 43
turtles, 191
    soft-shell, 220

vampyre squid, 257
velvet spider, **202**, 203–204, **205**, 206, 207
    body plan, 206
    DNA, 207
    eggs, 206
    evolution, 203
    eyes, 204–205
    genome, 203
    legs, 205–206
    mouth parts, 204
    sensory hairs, 204
    silk, 207
    thorax, 206
    venom, 204
    web shape, 207
Venerable Orang-outang, A (caricature), 15
vent shrimp, **144**, 145
    genome, 145
    lack of eyes, 145
Venter, Craig, 72
vertebrates, 72

Voytek, Mary, 72

Wallace, Alfred Russel, 14, 51, 87, 167, 175, 211, 252
whales, 33, 34–41
whaling industry, 41
whole genomics, 44–45
wolves, 115–117, **118**
    humans and, 118
woolly mammoth, **243**, 244–245
    body covering, 245
    brains, 144
    evolution, 244
    size, 245
    trunk, 244
    tusks, 244
    vs. African elephant, 243–245
Wroblewski, Emily, 188

Xenarthrans, 181

Yellowstone National Park, 96

zebra finch, **168**, 169, **170**, 171
    evolution, 169
    genes, 171
    longevity, 171
    red color, 170–171
    sex determination, 170
    small genome, 169
    talking gene, 171
zebrafish, **162**, **163**
    evolution, 163
Zoology of the Voyage of H.M.S. Beagle…During the Years of 1832–1836, The, (Darwin), 87

ABOUT THE AUTHOR

MAGGIE RYAN SANDFORD is a science journalist, broadcast media producer, science education researcher, and performer. Her work has appeared in magazines, museums, theaters, radios, and TVs around the world. She thinks you'd be really good at science and should give it a try.